D0982718

SAFETY SHUTDOWN SYSTEMS:
Design, Analysis, and Justification

By Paul Gruhn
and
Harry Cheddie

ISA is the international society for measurement and control

Copyright © 1998 by   Instrument Society of America
                         67 Alexander Drive
                         P.O. Box 12277
                         Research Triangle Park, NC   27709

Printed in the United States of America.
10 9 8 7 6 5 4 3 2

ISBN 1-55617-665-1

Library of Congress Cataloging-in-Publication Data

Gruhn, Paul.
      Safety shudown systems : design, analysis, and justification /
   Paul Gruhn and Harry Cheddie
            p. cm.
      Includes bibliographical references and index.
      ISBN 1-55617-665-1
      1. Chemical process control – Safety measures.   I. Cheddie, Harry.
   II. Title.
   TP155.75.G78   1998   98-38897
   660' .2815—dc21   CIP

TABLE OF CONTENTS

ABOUT THE AUTHORS

Harry L. Cheddie, P.Eng.

Harry Cheddie is an Engineering Associate with Bayer Inc. in Sarnia, Ontario, Canada, where he is mainly responsible for process control design and engineering.

Harry is a senior member of ISA and the American Society for Quality.

He teaches a course on the design and application of safety systems.

Harry graduated from Salford University in the UK with a B.Sc. (1st Class Honors) degree in Electrical Engineering. He is a registered Professional Engineer in the province of Ontario, Canada.

Harry is also certified by the American Society for Quality as a Quality Engineer, and as a Reliability Engineer.

Paul Gruhn, P.E.

Paul Gruhn is the owner of L&M Engineering and a Safety System Specialist with Moore Process Automation Solutions.

Paul is a senior member of ISA. He serves on the national Chemical & Petroleum Industries Division (CHEMPID) board as the Safety System Subcommittee Chairman, and sits on the SP84 Committee which wrote "Application of Safety Instrumented Systems for the Process Industries". Paul is the developer and instructor for ISA's 2-day course EC50, "Emergency Shutdown Systems". He is also a member of the American Society of Safety Engineers, the System Safety Society, the American Society of Quality, and the American Institute of Chemical Engineers.

Paul has a B.S. degree in Mechanical Engineering from Illinois Institute of Technology, in Chicago, Illinois and is a licensed Professional Engineer in Texas.

"Engineering responsibility should not require the stimulation that comes in the wake of catastrophe."

— S. C. Florman

What Is a "Safety Instrumented System"?

Safety interlock system, safety instrumented system, safety shutdown system, emergency shutdown system—the variety of names seems limitless. Within the process industry, debate continues over just what these systems should be called. Within the ISA SP84 committee there was continual discussion (and constant change) over the term used to describe these systems. The most generic term might be *safety system*, but this means different things to different people. For many chemical engineers, "safety systems" refer to management procedures and practices, not control systems. Many prefer the term *emergency shutdown systems*—ESD systems for short—but to electrical engineers ESD means "electrostatic discharge." Many don't want the word *emergency* appearing in the name at all because of its negative connotation. Others don't like the term *safety shutdown sys-*

tems for the same reason; anything appearing in print with the word *safety* in it draws immediate attention.

When the American Institute of Chemical Engineers (AIChE) published its guidelines text on this subject, the term it used was *safety interlock systems*. However, some would argue that interlocks are only one subset of many different safety control systems. The ISA SP84 committee finally settled on "safety instrumented systems" (SIS) so they could keep the same acronym used in the AIChE text SIS.

So just what *is* a safety instrumented system? Safety instrumented systems are designed to respond to conditions of a plant that may be hazardous in themselves or if no action were taken could eventually give rise to a hazard. They must generate the correct outputs to prevent the hazard or mitigate the consequences. This definition is a slight modification of the one presented in the U. K. HSE text *Programmable Electronic Systems in Safety-Related Applications*.

Who This Book Is For

This book is primarily intended for the thousands of instrumentation and control system engineers in the process industries (e.g., refining, chemical, petrochemical, offshore, etc.) who are responsible for designing, installing, and maintaining safety instrumented systems. These individuals are employed by end users, engineering firms, system integrators, and consultants. Managers and sales individuals will also benefit from a basic understanding of the material presented.

Why This Book Was Written

Today, we are engineering industrial processes–and using computers to control them–that have the potential for large-scale destruction. Even a single accident could be disastrous. We do not have the luxury of learning from experience ("Oops, we blew up that unit. Let's try lowering the set point five degrees"). We must attempt to anticipate and prevent accidents before they occur. Hopefully, this book will help.

We believe this to be the only all-encompassing text on this subject. This book is a practical "how-to" text on the analysis, design, application, and installation of safety instrumented systems. It includes practical knowledge required by engineers to apply safety instrumented systems. We hope it will serve as a guide for implementing the procedures outlined in the ANSI/ISA standard S84. It covers background material on risk assessment and management, the difference between process control and safety control, protection layers, human interfaces, field sensors, and actuators. The book focuses on such aspects of safety instrumented systems as design, analysis techniques, technology choices, purchasing, installation, documentation, and testing. Also covered are the technical and economic justifications for these systems. The focus throughout is on real-world, practical solutions with many actual examples and a minimum of theory and math. The equations that are presented involve only simple algebra.

Confusion in the Industry

One goal of this book is to clarify the general confusion in the industry regarding the design of these systems. Many would have hoped to turn to industry standards for their recommendations. Unfortunately, until recently, there have been few standards available to turn to.

Technology Choices

What technology should be used: relay, solid state, or microprocessor? Does this depend upon the application? Relay systems are still common for small applications, but would you want to design and wire a 500 I/O (input/output) system with relays? Is it economical to do a 20 I/O system using a redundant programmable system? Some people prefer not to use software-based systems in safety applications at all; others have no such qualms. Are some people "right" and others "wrong"?

Redundancy Choices

How redundant, if at all, should a safety instrumented system be? Does this depend upon the technology or the level of risk? If most relay systems were simplex (nonredundant), then why have triplicated programmable systems become so popular?

Field Devices

What about the field devices (sensors and valves)? Should sensors be switches or analog transmitters? Should "smart" (intelligent, i.e., processor-based) valves be used? Should field devices be redundant? How often should they be tested? Does this depend upon the application, the technology, and the level of risk?

Test Intervals

How often should systems be tested? Once per month, per quarter, per year, or per turnaround? Does this depend upon technology, level of redundancy, and level of risk? How does one even *make* all these decisions?!

Conflicting Vendor Stories

Every vendor seems to be touting a different story line, implying that only *their* system should be used. Who should one believe and, more importantly, *why*? How can one peer past all of the sales hype? When overwhelmed with choices, it becomes difficult to decide at all. Perhaps it's easier just to ask a trusted colleague what he or she did. Many people admit to being a bit confused.

Industry Standards

"Regulations are for the obedience of fools and for the guidance of wise men."

— RAF motto

One of the reasons industry writes its own standards, recommended practices, and guidelines is to avoid government regulation. If industry is responsible for accidents and decides not to regulate itself, it knows the government may step in and do it for them. Governments usually step in and get involved once risks are perceived by the general populace to be "alarming." The first successful regulatory legislation in the United States was passed by Congress over one hundred years ago, after public pressure and a series of marine steamboat boiler disasters killed thousands of people.

Health and Safety Executive: *Programmable Electronic Systems*

Programmable Electronic Systems in Safety-Related Applications, parts 1 and 2, U.K. Health and Safety Executive, 1987; ISBN 011-883913-6 and 011-883906-3.

This document, published by the English Health and Safety Executive, was the first of its kind. Although it focuses on software programmable systems, the concepts presented apply to other technologies as well. Available in the United States, it is an excellent document and covers qualitative and quantitative evaluation methods and many design checklists. It has been used as the foundation for several other documents.

American Institute of Chemical Engineers, Center for Chemical Process Safety

Guidelines for Safe Automation of Chemical Processes, AIChE, 0-8169-0554-1, 1993.

The American Institute of Chemical Engineers (AIChE) formed the Center for Chemical Process Safety after the accident in Bhopal, India, in 1984. It has since released about two dozen textbooks on various design and safety-related topics for the process industry. This particular AIChE text covers the design of DCS (distributed control systems) and interlock systems and contains other very useful background information. The book took several years to write and was the result of the effort of about a dozen individuals, all from user companies (i.e., no vendors).

IEC Draft Standard 61508

Functional Safety: Safety-Related Systems, IEC draft standard 61508, Technical Committee 65, Working Groups 9 and 10, 1997.

The International Electrotechnical Commission (IEC) has been working for years on a standard that covers the use of relay, solid-state, and programmable systems, including field devices. The standard applies for *all* industries, such as transportation, medical, nuclear, and so on. It is a seven-part document. Portions were approved for final vote in 1997.

ANSI/ISA S84.01-1996

Application of Safety-Instrumented Systems for the Process Industries, ISA standard S84.01-1996, 1996.

The ISA SP84 (standards and practices committee number 84) worked for over ten years developing this standard. The scope of this document has undergone many changes throughout the years. It was originally intended as a U.S. standard and was only going to address the logic box (and not the field devices). Then the IEC group started on its general document, and the SP84 group felt its document could be used as an industry-specific document (for the process industries) under the scope of the IEC. This eventually happened, and in the process the SP84 group included field devices and other technologies within its scope. The document has undergone major changes over the years. Unfortunately, the long-term future of S84 is unclear because the IEC has formed its own task force to write the IEC process industry-specific document (IEC draft standard 61511). ANSI/ISA S84.01-1996 was released in early 1996, but what happens to it after the IEC 61511 document is released is uncertain. The American

National Standards Institute (ANSI) is the U.S. representative within the IEC. In all likelihood, ANSI will support the IEC 61511 standard when it is released and replace ANSI/ISA S84.01-1996.

OSHA (29 CFR 1910.119)

OSHA, 29 CFR 1910.119, *Process Safety Management of Highly Hazardous Chemicals.*

The process industry has a vested interest in writing its own industry standards, guidelines, and recommended practices. As stated earlier, if industry was seen as being unable to control its own risks, there would be the possibility of government intervention. This has happened with the release of process safety management legislation. These laws are directed at companies dealing with highly hazardous substances. There are over a dozen sections to this legislation. A number of the sections have requirements specifically calling out issues related to the selection, design, documentation, and testing of safety instrumented systems.

For example:

Section d3, Process safety information: "Information pertaining to the equipment in the process … [including] safety systems … 'For existing equipment … the employer shall *determine and document* that the equipment is designed, maintained, inspected, tested, and operating in a *safe manner'* " (emphasis added).

People tend to have more questions *after* reading the OSHA document than before. For example, just what is "a safe manner"? How does one "determine" and in what way does one "document" that things are operating "safely"? The OSHA document does little to answer these questions.

Section j, Mechanical integrity: "Applies to the following process equipment: … , emergency shutdown systems… *Inspection and testing:* 'The frequency of inspections and tests of process equipment shall be consistent with applicable manufacturer's recommendations and good engineering practices, and more frequently if determined to be necessary by prior operating experience.' " Whose experience?

Section j5, Equipment deficiencies: "The employer shall correct *deficiencies* in equipment that are outside *acceptable limits* before further use or in a safe and timely manner when necessary means are taken to *assure safe operation*" (emphasis added). What is the definition of a "deficiency"? This sentence would seem to contradict itself. It first introduces the idea of "acceptable limits": if I stand here, it's acceptable, but if I step over an imaginary boundary and stand over there, it's no longer acceptable. This seems harmless enough. But the very same sentence then goes on to say that if anything goes wrong, you obviously didn't "assure" (guarantee) safe operation. In other words, no matter what happens, you *can't win.*

Section j6, Quality assurance: "In the construction of new plants and equipment, the employer shall *assure* that equipment as it is fabricated is *suitable* for the process application for which they will be used" (emphasis added). The employer shall "assure"? Benjamin Franklin said the only thing we can be "sure" of is death and taxes. "Suitable"? According to who? The vendor trying to sell you its system? Measured against what?

Appendix C, Compliance guidelines and recommendations, Mechanical integrity: "Mean time to failure of various instrumentation and equipment parts would be known from the manufacturer's data or the employer's experience with the parts, which would then influence the inspection and testing frequency and associated procedures." Are companies aware that they are supposed to be keeping records of this sort of information? Just how would this "influence" the test frequency of various systems? How does one even make this determination? Some manufacturers have and do provide failure rate data—some do not.

Standards Are Changing Their Direction

Most people want a simple "cookbook" of preplanned solutions. For example, for a refinery, simply "turn to page 35." There it shows dual-sensor, dual-logic, simplex valves, yearly test interval, and the like. For an offshore platform, "turn to page 63." There it shows ... and so on. There are reasons why the standards will never be written this way. The standards do *not* give clear, simple, precise answers. They do *not* mandate technology, level or redundancy, or test intervals.

There is a fundamental change in the way industry standards are being written. Standards are moving away from *prescriptive* standards and toward more *performance*-oriented requirements. This means each plant will have to decide for itself just what is "safe," and each plant will have to decide how it will "determine" and "document" that its systems are, in fact, safe. Unfortunately, these are difficult decisions that few want to make, and fewer still want to put in writing. "What is safe" transcends pure science and involves philosophical and moral matters.

Things Are Not As Obvious As They May Seem

Intuition and gut feelings do not always lead to correct conclusions. For example, which system is safer, a dual one-out-of-two (1oo2) system (where only one of the two redundant channels is required in order to generate a shutdown) or a triplicated two-out-of-three (2oo3) system (where two of the three redundant channels are required in order to generate a shutdown)? Intuition might lead you to believe that if one system is "good," two must be better, and three must be the best. So you might conclude that the triplicated system is best. Unfortunately, it's not. The dual system is safer.

If there hasn't been an accident in your plant for the past fifteen years, does that mean you have a safe plant? It might be tempting to think so, but nothing could be further from the truth. Just because it hasn't happened yet doesn't mean it won't or can't. No doubt people might very well have made such a statement the day before Seveso (Italy), Flixborough (England), Bhopal (India), Chernobyl (Soviet Union), Pasadena (United States), and so on.

If design decisions regarding safety instrumented systems were simple, obvious, and intuitive, there would be no need for this book. Airplanes and nuclear power plants are not designed by intuition or gut feeling. How secure and safe would you feel if you asked the chief engineer of the Boeing 777 "Why did you choose that size engine?" and his or her response was, "That's a good question. We really weren't sure, but that's what our vendor recommended." You'd like to think that Boeing would know how to engineer the entire system (and indeed they do). Why should safety instrumented systems be any different?

Many of the terms used to describe safety instrumented system performance seem so simple and intuitive, yet they can cause so much confusion. For example, can a system that is ten times more reliable be less safe? If we were to replace a relay-based shutdown system with a newer PLC (programmable logic controller) that the vendor said was ten times more reliable than the relay system, would it automatically follow that the system was safer as well? Safety and reliability are not necessarily the same thing. It is actually very easy to show that one system may be more reliable than another but may still be *less safe*.

The Danger of Complacency

It is easy to become overconfident and complacent about safety. It is easy to believe that we as engineers using modern technology can overcome almost any problem. History has proven, however, that we cause our own problems, and we always have more to learn. Bridges will occasionally fall, planes will occasionally crash, and petrochemical plants will occasionally explode. That does not mean, however, that technology is bad or that we should live in the Stone Age. It's true that cavemen didn't have to worry about The Bomb, but then we don't have to worry about the plague.

After Three Mile Island (the worst U.S. nuclear incident), but before Chernobyl (the worst ever nuclear incident), the head of the Soviet Academy of Sciences said, "Soviet reactors will soon be so safe that they could be installed in Red Square." Do you think he'd say that *now*?

The works manager at Bhopal was not in the plant when that accident happened. When he was finally located, he could not accept that his plant was actually responsible. He was quoted as saying, "The gas leak just can't be from my plant. The plant is shut down. Our technology just can't go wrong. We just can't have leaks." One wonders what he does for a living now.

Technology may be a good thing, but it is not infallible. We as engineers and designers must never be complacent about safety.

So now that we've raised some of the questions, let's see how to answer them.

1

DESIGN LIFE CYCLE

"If I had 8 hours to cut down a tree, I'd spend 6 hours sharpening the axe."

— *A. Lincoln*

Designing a single component may be viewed as a relatively simple matter, one that a single person can handle. Designing any large system, whether it's a car, a computer, or an airplane, however, is typically beyond the ability of any one individual. Large systems require a multidisciplinary *team*. The control system engineer should *not* feel that the entire burden of designing a safe plant rests on his or her shoulders alone because it obviously does not.

1.1   Hindsight/Foresight

"Hindsight can be valuable when it leads to new foresight."
                                                    — *P. G. Neumann*

Hindsight is easy. Everyone always has twenty-twenty hindsight. Foresight, however, is a bit more difficult. Foresight is required, however, with today's large, high-risk systems. We cannot afford to design large petrochemical plants by trial and error; the risks are simply too great. We have to try and prevent certain accidents, no matter how remote their possibility, even if they have never yet happened. This is the goal of *system safety*.

System safety was born out of the needs of the aerospace industry. The military offers obvious examples of high-risk systems. The following case, for example, may have been written in a lighthearted vein, but it was obviously a very serious matter to the personnel involved (luckily, there were no injuries):

An ICBM silo was destroyed because the counterweights, used to balance the silo elevator on the way up and down, were designed with consideration only to raising a fueled missile to the surface for firing. There was no consideration that, when you were not firing in anger, you had to bring the fueled missile back down to defuel. The first operation with a fueled missile was nearly successful. The drive mechanism held it for all but the last five feet when gravity took over and the missile dropped back down. Very suddenly, the 40-foot diameter silo was altered to about a 100-foot diameter.[1]

Similarly, a radar warning system in Greenland suffered an operational failure in its first month. It reported inbound Russian missiles, but what it actually responded to was ... *the rising moon.*

If you make something available to someone, it will at some point be used, even if you didn't intend it to be. For example, there were two cases where NORAD (North American Air Defense Command) and SAC (Strategic Air Command) went on alert because radar systems reported incoming missiles. In reality, someone just loaded a training tape by mistake. After the same accident happened a *second* time, training tapes were stored in a different location. What might have originally been considered "human error" was really an error in the "system" that allowed the (inevitable) human error to happen.

1.2 Findings of the Health and Safety Executive

The English Health and Safety Executive (HSE) examined thirty-four accidents that were the direct result of control and safety system failures in a variety of different industries.[2] Their findings are summarized in Figure 1-1. A majority of accidents were caused by incorrect specifications (both the functional specification as well as the integrity specification). Another large portion was caused by changes made after commissioning. These two areas alone accounted for 64 percent of the problems.

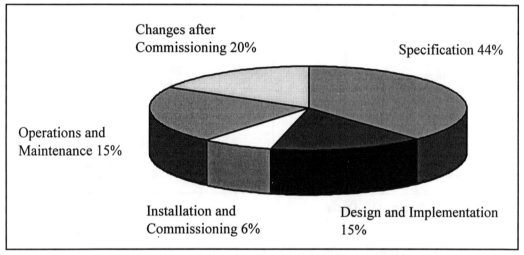

Figure 1-1   Accident Cause by Phase As Found by the United Kingdom's HSE

Nancy Leveson has reported that the vast majority of accidents in which software was involved could be traced back to flaws in the requirements.[3] In other words, incomplete or wrong assumptions about the operation of the controlled system or required operation of the computer, unhandled system states and environmental conditions. To design a "safe" system, one needs to consider many different areas and not just focus on the one that may be the easiest to cover. The experts formulating the various industry standards and guidelines realize this and are trying to cover all of the bases. This text will attempt to do the same.

1.3 Design Life Cycle

Large systems require a methodical design *process* to prevent important items from falling through the cracks. Figure 1-2 shows the life cycle steps as described in the ANSI/ISA S84.01-1996 standard. This should be considered as only one example; variations of the life cycle are presented in other industry documents as well. A company may wish to develop its own variation of the life cycle based upon its unique requirements.

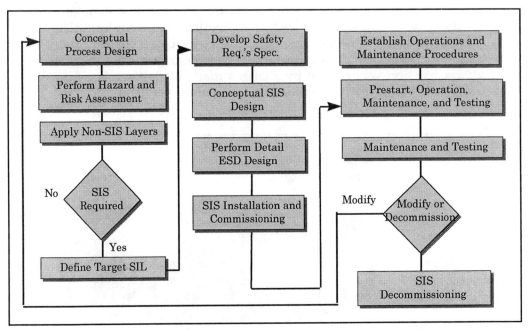

Figure 1-2   Design Life Cycle

Some will complain that performing all of the life cycle steps, like all other tasks designed to lower risk, will increase overall costs and result in lower productivity. Several years ago, one in-depth study conducted by a group including major engineering societies, twenty industries, and sixty product groups with a combined exposure of over fifty billion hours concluded that *production increased as safety increased.*[3]

1.3.1   Conceptual Process Design

The first step in the life cycle is to develop an understanding of the process, the equipment under control, and the environment (physical, social, political, and legal) in sufficient depth to enable the other life cycle activities to be performed. The goal is to design an inherently safe plant. The activities in this step are generally considered outside the realm of the control system engineer.

1.3.2 Hazard Analysis and Assessment

The next step is to develop an understanding of the risks associated with the process. Risks may impact personnel, production, capital equipment, the environment, company image, and the like. *Hazard analysis* consists of *identifying* the hazards. There are numerous techniques one can use (HAZOP, what-if, fault tree, checklist, etc.) and numerous texts describing each method. *Risk assessment* consists of *ranking* the risk of the hazards that have been identified in the hazard analysis. Risk is a function of the frequency or probability of an event and the severity or consequences of the event. Risk assessment can be either qualitative or quantitative. Qualitative assessments subjectively rank the risks from low to high; quantitative assessments, as their name obviously implies, attempt to assign numerical factors to the risk, such as death or accident rates. This is not intended to be the sole responsibility of the control system engineer. Obviously, a number of other disciplines are required in order to perform these assessments.

1.3.3 Application of Non-Safety-Instrumented- System Layers

The goal of process plant design is to have a plant that is inherently safe, one where residual risks can be controlled by the application of noninstrumented safety layers. KISS (Keep It Simple, Stupid) should be an overriding theme.

1.3.4 Is a Safety Instrumented System Required?

If the risks can be controlled to an acceptable level without the application of an instrumented system, then the design process stops (as far as a safety instrumented system [SIS] is concerned). If the risks cannot be controlled to an acceptable level by the application of noninstrumented layers, then an instrumented system will be required.

1.3.5 Define Target Safety Integrity Level

Safety instrumented system (SIS) performance should match the level of risk. In other words, the greater the level of process risk, the better the SIS needs to be in order to contain the risk. This requires identifying the individual risks and assessing their impact.

The most difficult step in the overall process for most organizations seems to be determining the required safety integrity level (SIL). This is not directly a measure of process risk but rather a measure of the safety system performance that is needed to control the risks identified earlier to an acceptable level. The standards describe qualitative methods regarding how this can be done. A potential problem with the qualitative methods is that they are subjective and very often not repeatable. Different organizations may review the same process and each come up with different SIL requirements. There are also quantitative methods available for determining system performance. A problem for many trying to use quantitative methods, however, is that they must decide on a quantitative target safety goal. Deciding "what is a tolerable death rate" is something few people wish to do and fewer companies wish to put in writing. One can just imagine an attorney saying, "What do you *mean* you considered it *tolerable* to *kill* four people every 100 million man-hours?!"

A mechanical analogy may be helpful. Suppose we have a requirement for a valve. The valve must be a certain size, must close within a certain period of time, must operate with a certain pressure drop, and so on. This represents the functional specification—*what* the valve is to do. If a carbon steel valve is installed and two months later the valve is found to have failed because of heavy corrosion where was the error made? The functional specification just described *what* the valve was to do. In this case, the integrity specification was incomplete and did not state the corrosive service and that a special material was required. The integrity specification states how *well* the system should perform its function. In terms of an SIS, this would be the SIL.

1.3.6   Develop Safety Requirements Specification

The next step consists of developing the safety requirements specification, essentially the functional logic of the system. This will naturally vary for each system. No general, across-the-board recommendation can be made. For example, if temperature sensor TT2301 exceeds 410°F, then close valves XV5301 and XV5302. Each safety function should have an associated SIL requirement as well as reliability requirements if nuisance trips are a concern. One should include *all* operating conditions of the process, from start-up through shutdown, as well as maintenance. One may find that certain logic conditions conflict during different operating modes of the process.

The system will be programmed and tested according to the logic determined during this step. If an error is made here, it will carry through for the rest of the design. It won't matter how redundant or how often the system is manually tested—it will not work properly when required. These are referred to as *systematic* or *functional failures*. Using diverse redundant systems programmed by different people using different languages and tested by an independent team will not help in this situation because the functional logic they all based their work on could be in error.

1.3.7 Conceptual SIS Design

One doesn't pick a certain size jet engine for an aircraft based on intuition. One doesn't size a million-dollar compressor by "gut feeling." One doesn't determine the size of pilings required for a bridge by trail and error (at least not any more).

The purpose of the conceptual SIS design step is to develop an initial design to see if it meets the safety requirements and SIL performance requirements. One needs to initially select a technology, configuration (architecture), test interval, and so on. This pertains to the field devices as well as the logic box. Factors to consider are overall size, budget, complexity, speed of response, communication requirements, interface requirements, method of implementing bypasses, testing, and the like. One can then perform a quantitative analysis to see if the proposed system meets the performance requirements or make a qualitative judgment based on prior experience (although this is obviously harder to substantiate). The intent is to evaluate the system *before* one specifies the solution. Just as it is better to perform a HAZOP *before* you build the plant (it's hard to change the design once it's already been built), it is better to analyze the proposed safety system *before* you specify it, or else how will you know if it meets the performance goal?

1.3.8 Detailed SIS Design

The purpose of the detailed SIS design step is to finalize and document the design. Once a design has been chosen, the system must be engineered and built following strict and conservative procedures. This is the only realistic method for preventing design and implementation errors that we know of. The process requires thorough documentation, that is, an auditable trail that someone else may follow for verification purposes.

1.3.9 Installation and Commissioning

The installation and commissioning step is to ensure that the system is installed according to the design and performs according to the safety requirements specification. Before a system is shipped from a factory, it must be thoroughly tested for proper operation. If any changes are required, they should be made at the factory, not at the installation site. At installation, the entire system, this time including the field devices, must be checked as well. There should be a detailed installation document outlining each procedure to be carried out. Finished operations should be signed off in writing to show that each function and operational step has been checked.

1.3.10   Operations and Maintenance

To function properly, every system requires periodic maintenance. Not all faults are self-revealing, so *every* SIS *must* be periodically tested to make sure it will respond properly to an actual demand. The frequency of inspection and testing will have been determined earlier in the life cycle. All testing must be documented.

1.3.11   Modifications

As process conditions change, it will be necessary to make changes to the safety system. All proposed changes require that you return to the appropriate phase of the life cycle. A change that may be considered minor by one individual may actually have a major impact on the overall process. This can only be realized if the change is thoroughly reviewed by a qualified team. Hindsight has shown that many accidents have been caused by this lack of review.

1.3.12   Decommissioning

Decommissioning a system should entail a review to make sure that removing the system from service will not impact the process or surrounding units and that the means are available during the decommissioning process to protect the personnel, equipment, and environment.

Summary

The overall design of SISs is not a simple, straightforward matter. The engineering skills required in the design of such systems are often beyond that of the control systems engineer. An understanding of the process, instrumentation, control systems, and hazard analysis are required. This typically requires the interaction of a multidisciplinary team.

Experience has shown that a detailed, systematic, methodical, well-documented design *process* is called for in the design of SISs. This starts with a safety review of the process, then the implementation of other safety layers followed by systematic analysis and detailed documentation and procedures. The steps are described in most documents as a *safety design life cycle*. The intent is to leave a documented, auditable trail and make sure that nothing falls between the inevitable cracks in any organization and is neglected.

References

1. *Air Force Space Division Handbook.*

2. U.K. Health and Safety Executive, *Out of Control: Why Control Systems Go Wrong and How to Prevent Failure* U.K. Health and Safety Executive, 1995), ISBN 0-7176-0847-6.

3. Nancy G. Leveson, *Safeware: System Safety and Computers* (Reading, MA: Addison-Wesley, 1995), ISBN 0-201-11972-2.

4. ANSI/ISA, *Application of Safety Instrumented Systems for the Process Industries*, ANSI/ISA S84.01-1996 (Research Triangle Park, NC: ISA), ISBN 1-55617-590-6.

5. International Electrotechnical Commission, *Functional Safety: Safety-Related Systems*, draft standard 61508 (Geneva, Switzerland: IEC, 1997).

6. U.K. Health and Safety Executive, *Programmable Electronic Systems in Safety-Related Applications, Part 1: An Introductory Guide* (Sheffield, UK: U.K. Health and Safety Executive, 1987), ISBN 011 883913 6.

7. American Institute of Chemical Engineers, *Guidelines for Safe Automation of Chemical Processes* (New York: American Institute of Chemical Engineers, Center for Chemical Process Safety, 1993), ISBN 0-8169-0554-1.

2

RISK

Now let me get this straight. You smoke two packs a day, but still wear seat belts?

Gruhn

"Whether quantitative or qualitative, risk assessment is much like statistics; it can demonstrate almost any conclusion, depending on the assumptions and the frame of reference."

— *P. Neumann*

Risks are everywhere. Working in a chemical plant obviously involves risk, but then so does taking a bath. One involves more risk than the other, but there is a measure of risk in everything we do.

Although zero injuries may be a goal for many, it is important to realize that there is no such thing as zero risk. One is at risk merely sitting at home watching television. So how safe should a process plant be? Should the risk of working at a chemical plant be equal to that of staying at home or driving one's car or flying in an airplane—or skydiving? All things being equal, the safer a plant is, the more expensive it will be. There has to be an economic consideration at some point. If the safety goal prevents the process from even being started, something obviously needs to be changed.

2.1 Hazard

According to the American Institute of Chemical Engineers (AIChE) the definition of a hazard is "an inherent physical or chemical characteristic that has the potential for causing harm to people, property, or the environment. In this document, it is the combination of a hazardous material, an operating environment, and certain unplanned events that could result in an accident."

2.2 Risk

Risk is usually defined as the combination of the severity and probability of an event. In other words, how often can it happen and how bad is it when it does? Risk can be evaluated qualitatively or quantitatively.

People are not the only thing at risk in process plants. Unmanned platforms have a considerable amount of risk, even though there may be no personnel on the platform most of the time. Also at risk are uninterrupted production uptime, capital equipment, the environment, the company's budget for litigation costs, and company image. Some of these items (e.g., lost production, capital equipment) may be quantified; some others (e.g., company image) may be more difficult to put a number on.

2.3 As Low As Reasonably Practical (ALARP)

There are three general levels of risk. At the one extreme are negligible risks—the ones that are so low as to not be of concern. A simple example would be the risk of being struck by lightning. (Actually, about 150 people in the United States die this way every year, but it is doubtful that anyone loses sleep worrying about the possibility.) At the other extreme are risks that are viewed as being so high as to be considered unacceptable. An example in the United States might be nuclear reactors. (No new reactors have been installed in the United States for over a decade.) In between these two extremes are risks that are considered tolerable, or acceptable, *if* the positive benefit is seen to outweigh the potential negative impact. Driving a car is an excellent example. The fact that approximately forty-five thousand people in the United States die in auto accidents every year does little to deter the rest of us from driving. These three levels of risk are shown in Figure 2-1.[3]

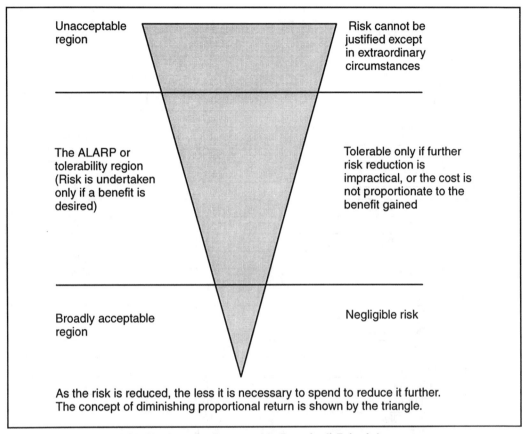

Figure 2-1   ALARP (As Low As Reasonably Practical) Principle

2.4 Fatality Rates

There are two common methods of expressing fatality rates. One is the fatal accident rate, or FAR. It is the number of deaths per million person-hours of exposure. Another is the probability per unit of time. A number of sources list fatality rates for various activities, for different industries, means of transportation, and recreational activities, including voluntary as well as involuntary risks. Table 2-1 lists fatality rates for various activities in the United Kingdom.[1] Note that probabilities for two activities may be the same, yet the FAR may be different because of the difference in exposure hours. Also, note that the numbers come from different sources, and all of the underlying information may not be available. Consequently, there is not always a means of directly relating probabilities to rates, hence some of the fields in Table 2-1 are blank.

Activity	Probability (per year)	FAR
Travel		
Air	2×10^{-6}	
Train	3×10^{-6}	3 - 5
Bus	2×10^{-4}	4
Car	2×10^{-4}	50-60
Motorcycle	2×10^{-2}	500 - 1,000
Occupation		
Chemical Industry	5×10^{-5}	4
Manufacturing		8
Shipping	9×10^{-4}	8
Coal Mining	2×10^{-4}	10
Agriculture		10
Boxing		20,000
Voluntary		
The Pill	2×10^{-5}	
Rock Climbing	1.4×10^{-4}	4,000
Smoking	5×10^{-3}	
Involuntary		
Meteorite	6×10^{-11}	
Falling Aircraft	2×10^{-8}	
Natural Disasters	2×10^{-6}	
Firearms	1×10^{-6}	
Cancer	2.5×10^{-5}	
Fire	2×10^{-5}	
Falls	1×10^{-4}	
Staying at Home		1 - 4

Table 6-1   Fatal Accident Rates (in the United Kingdom)

2.5   Risks Inherent in Modern Society

Just how safe is "safe"? Should working in a plant have the same level of risk as skydiving (which kills about forty people per year in the United States)? Should working in a plant be as safe as driving your car? (Relatively speaking, of course. We certainly can't afford to kill forty-five thousand people in the U.S. process industry every year.) Or should it be as safe as flying in a plane, which is safer than driving a car by two orders of magnitude?

Although the term FAR may be simple to understand and may represent a useful "yardstick," it is doubtful that U.S. companies will ever put such "targets" in writing. Imagine walking into company XYZ's plush world headquarters and on the wall in the reception area is a sign that reads, "We at XYZ consider it tolerable to kill four people per 100 million man-hours." The lawyers would have a field day!

People's perception of risk varies depending upon their understanding or familiarity with the risk. For example, most people are familiar with driving. The perceived level of risk is relatively low to most people, even though, as noted, approximately forty-five thousand people die in the United States annually alone due to traffic accidents. If a new chemical facility is being proposed near a residential area, the level of understanding of the residents regarding the chemical process will probably be low, and their discomfort, or perceived level of risk, will no doubt be high, even if the process may have a very good historical safety record.

Perception of risk will also vary in proportion to the number of possible deaths associated with a particular event. For example, of the forty-five thousand traffic accident deaths every year, the deaths usually occur one, or a few, at a time. Even with this surprisingly high number of deaths, there is little (if any) public outcry that something be done to lower this figure. Yet when there is an accident with a school bus involving injuries to many children, there typically *is* an outcry. The same could be said about relatively high-risk sports, such as skydiving, hang gliding, and ultralight aircraft. Although these sports involve relatively high risk, it is rare that one hears of multiple deaths. Accidents involving the chemical industry, however, frequently do involve multiple deaths. (Bhopal was the worst, with over two thousand deaths and two hundred thousand injuries.) Even though the overall risk between driving and working in a chemical plant is about the same, and the risk is voluntary for the workers in the plant (some may argue this point), the public's perception of the riskiness of the two activities is much different.

2.6   Voluntary versus Involuntary Risk

There is a difference between voluntary and involuntary risks. Examples of voluntary risks would be driving a car, smoking cigarettes, and so on. Examples of involuntary risks would be having a chemical plant built near your home *after* you have lived there a number of years or secondary smoke from other peoples' cigarettes.

People can perceive similar risks differently. For example, the Jones's own a house in the country. One day, company XYZ builds a toxic chemical plant nearby. After the plant is built, the Smiths buy a house next to the Jones's. Both households face the same risk, but they will probably each have a different perception of it. To the first couple (the Jones's) who lived there *before* the plant was built, the risk is involuntary (although they obviously could move). To the second couple (the Smiths), who bought their home *after* the plant was built, it's voluntary.

People are usually willing to face higher voluntary risks than involuntary ones. For example, when one of the authors was younger he was willing to accept both the risks of riding a motorcycle in a large city and of skydiving. The risks were voluntary, and he considered himself the only one at risk at the time. (One could argue the finer points of that.) As a married father, he no longer wishes to accept those risks (never mind the fact that he can no longer afford them.)

Another factor involved in the perception of risk is *control*. For example, the wife of one of the authors does not like flying. (In fact, flying is the number two fear among Americans. Public speaking is number one.) Her reason for discomfort is that she does not feel she is "in control." When you're sitting behind the wheel in your car at a stop sign and a drunk driver plows into your car, you weren't in control then either. No one goes out *planning* to have an accident.

2.7   Acceptable/Tolerable Levels of Risk

The concept of acceptable or tolerable levels of risk is not solely a technical issue; it involves philosophical and moral issues. Deciding how safe is safe enough cannot be answered by algebraic equations or probabilistic evaluations. Alvin Weinberg has termed these "transscientific questions," for they transcend science.[2]

Another issue that presents difficulties is trying to statistically estimate extremely unlikely events. Estimating such events, such as a severe chemical accident, cannot have the same validity as estimates for which abundant statistics are available. Because the required probabilities can be so small (e.g., 10^{-6} per plant per year), there is no practical means of determining the rate directly—that is, building ten thousand plants and operating them for one hundred years in order to tabulate their operating histories. Putting it a simpler way, measuring something that doesn't happen very often is difficult.

2.8   Acceptable Risk in the Process Industries

It is common to view personal risk in a subjective, intuitive manner. Many people will not consider driving a motorcycle, no matter what their biker friends may say. The wife of one of the authors who does not like flying believes it is all right for her family to drive to the airport in the same car but not to fly in the same airplane, even though she understands that flying is two orders of magnitude safer. Obviously, logic does not always apply when one is determining relative risk.

We should not, however, have the same subjective attitude about risks in the process industry. Usually, the people making the risk decisions (engineers) are not the ones who will be facing the eventual risk (nearby residents). Although none of the more famous accidents in the process industry would ever be considered "acceptable," the companies involved did *not* go out of business the next day. Therefore, the losses *must* have been considered "tolerable." How many accidents is the industry willing to consider tolerable, however, before there is public and political outcry?

There are approximately twenty-three hundred petrochemical plants in the United States alone. If we blew up one every year because of an accident (which represents an individual plant risk of 1/2,300 per year), how long would it be before the government stepped in? What if such an accident only happened once every ten years (1/23,000 per year)? There is no such thing as zero risk, but it is very difficult to decide what particular level should be considered "tolerable."

The relevant statistics can be rather confusing. An individual risk of one in twenty-three hundred per year means that out of twenty-three hundred plants, one might go "boom" every year, on average. It is important to realize, however, that you can't predict which plant, and you can't predict when one will go "boom." But since people don't build twenty-three hundred plants all at once, or live next to twenty-three hundred plants, they want to know the risk of *one* plant. The risk for an individual plant remains the same, 1/2,300 per year, but some are just not comfortable with such a number. Some confuse the number and say the risk of an accident is "once per twenty-three hundred years. This causes even more confusion. Some then assume it will be twenty-three hundred years before there is an accident. Nothing could be further from the truth.

Deciding what level of risk is acceptable could be compared to choosing your weapon in Russian roulette—how many barrels do you want in your gun? Would you choose an automatic pistol that always had a round in the chamber? (I hope not!) Or would you choose a revolver with six chambers? What if you could choose a gun with fifty barrels or one with five thousand barrels. Obviously, the more the barrels, the lower the risk of hitting the bullet when you pull the trigger.

Twenty-three hundred years between accidents may initially sound ridiculous. A more intuitive answer might initially be fifty years (the reasoning apparently being that someone's working life is fifty years, and they don't want anything to happen during *their* life, so ...). But how long will the plant *actually* be around? Let us say twenty-five years. So take a gun with fifty barrels, and pull the trigger once a year. What is the likelihood of firing the bullet? 50 percent. (Based on another set of assumptions and simplifications, however, the answer is 40 percent. Even the statistics involved are not inherently obvious.)

Would you want to have *your* name on the drawings of a plant if you knew there was a 50 percent chance of a catastrophic accident happening during the life of the plant? Probably not.

What if instead of fifty years, you choose five hundred years? Then the risk would now be 25/500, or a 5 percent chance. Should that be tolerable? What about five thousand years? Now it becomes 0.5 percent. There is no such thing as zero risk, but just how low must one go for the risk to be considered tolerable? That's the magical $64,000 question to which there is no scientific answer.

U.S. industry cannot afford to blow up a plant once every year (1/2,300). A serious accident once every ten years *might* be viewed as tolerable (1/ 23,000). Risk targets in the range of 1/10,000 have been documented. In fact, in The Netherlands, the government even publishes what it considers are tolerable fatality rates.

Summary

How safe should it be working at or living near a chemical plant? Should it be as safe as driving your car or flying in an airplane? These are questions that transcend science. Economics play a factor. At some point, it no longer becomes feasible to try and prevent certain events, hence the concept of ALARP (As Low As Reasonably Practical).

Risk to humans can be expressed as fatality rates or death probabilities per unit of time. Comparisons can then be made between different activities, occupations, methods of transportation, and so on. Risks are perceived differently by different people. Factors such as familiarity, level of perceived control, and number of fatalities per event enter into our perception of risk.

References

1. David J. Smith, *Reliability, Maintainability, and Risk (Practical Methods for Engineers)*, 4th ed. (Oxford: Butterworth-Heinemann, 1993), ISBN 0-7506-0854-4.

2. Nancy G. Leveson, *Safeware: System Safety and Computers* (Reading, MA: Addison-Wesley, 1995), ISBN 0-201-11972-2.

3. International Electrotechnical Commission, *Functional Safety: Safety-Related Systems*, draft standard 61508 (Geneva, Switzerland: IEC, 1997).

Additional Background Material

John Withers, *Major Industrial Hazards* (New York:Halsted Press, 1988), ISBN 0-470-21067-2.

J. R. Taylor, *Risk Analysis for Process Plant, Pipelines, and Transport* (London: E & FN Spon, 1994), ISBN 0-419-19090-2.

The Hon. Lord Cullen, *The Public Inquiry into the Piper Alpha Disaster* (London: Her Majesty's Stationery Office, 1990), ISBN 0-10-113102.

3

PROCESS CONTROL VERSUS SAFETY CONTROL

The keys are in the cabinet, there're jumpers where they shouldn't be, the password's taped to the monitor. I'd say we have a security problem.

Gruhn

In the past, process control used to be performed in pneumatic analog single loop controllers. Safety functions were performed in different hardware, typically hardwired relay systems. Electronic distributed control systems (DCSs) started to replace single loop controllers in the 1970s. Programmable logic controllers (PLCs) were developed to replace relays in the late 1960s. Since both systems are software programmable, some people naturally concluded there would be benefits from performing both the control and safety functions within the same system, usually the DCS. (The typical benefits touted included a single source of supply, integrated communications, reduced training and spares, and potentially lower overall costs.) Some argue that the reliability (and redundancy) of modern DCSs is "good enough" to allow such combined operation. All domestic and international standards, guidelines, and recommended practices, however, clearly recommend the *separation* of the two systems. The authors of this book agree and also recommend the separation of control and safety systems. We also wish to stress that *the "reliability" of the DCS is not the issue.*

3.1 Process Control: Active/Dynamic

It is important to realize and understand the fundamental difference between process control and safety control. Process control systems are active, or dynamic. They have analog inputs and analog outputs, perform math and number crunching, and have feedback loops. Hence, most failures in these systems are inherently self-revealing. For example, consider the case of a robot on an automated production line. Normally, the robot picks up part A and places it in area B. If the system fails, it's obvious to everyone–the robot no longer places part A in area B. There's no such thing as a "hidden" failure. The system either works or it doesn't. There is only one failure mode with such systems, and you don't need extensive diagnostics to annunciate such failures.

3.1.1 Need to Make (and Ease in Making) Frequent Changes

Process control systems must be flexible enough to allow frequent changes. Process parameters (e.g., set points, PID settings, manual/automatic, etc.) require changing. Portions of the system may also be placed in bypass, and the process may be controlled manually. This is normal for control systems.

3.2 Safety Control: Passive/Dormant

Safety systems, however, are just the opposite of process control systems. They are dormant, or passive. They sit there doing nothing and hopefully will never be called into action. An example would be a pressure relief valve. Normally, the valve is closed. It only opens when the pressure reaches a certain limit. If the pressure never exceeds that value, the valve never operates. Many failures in these systems may *not* be self-revealing. If the relief valve is plugged, there is no immediate indication. A PLC could be hung up in an endless loop. Without a watchdog timer, the system would not be able to recognize the problem. An output module might use triacs, which could fail energized. Many systems would not be able to recognize such a problem. How much confidence can you place in something that under normal circumstances is not functioning (or changing state)? The answer is you must test the system or it must be able to test itself. Hence, there is a need for extensive diagnostics in dormant, passive safety-related systems. (The alternative would be to use inherently fail-safe systems where such dangerous failures are highly unlikely.)

Systems designed for process control may be totally unsuitable for safety control. The fundamental difference is the level of self-diagnostics. If the system is "passive," failures are not self-revealing, hence extensive diagnostics are required. Many people assume that all modern electronics includes extensive diagnostics. Unfortunately, this is *not* the case. There are very simple economic reasons why this is so.

3.2.1   Defining a "Safety Critical" System

Critical systems require testing and thorough documentation. It is debatable whether normal process control systems require the same rigor in testing and documentation. When the U.S. government came out with the Process Safety Management (PSM) legislation (29 CFR 1910.119), many questioned whether the level of documentation and testing required addressed the control systems as well as the safety systems. For example, most organizations have documented testing procedures for their safety instrumented systems (SIS), but the same may not be said for all of their control system loops. Users in the process industry questioned OSHA representatives as to whether the requirements outlined in the PSM legislation applied to all six thousand loops in their DCS or just the two hundred in their safety instrumented systems. OSHA's natural response was to include everything. Users felt this was another nail in the proverbial coffin to put them out of business.

This helped spur the development of ANSI/ISA S91.[1] This brief standard (only two pages long) includes a definition of *safety critical*. Although some would argue with it, the standard's definition of *safety critical* is "a control whose failure to operate properly will directly result in a catastrophic release of toxic, reactive, flammable or explosive chemical." Based on this definition, users told OSHA that their six thousand DCS loops were not safety critical and therefore did not require the same degree of rigor for documentation and testing as their two hundred safety instrumented loops.

This is not meant to imply that the design of distributed control systems does not require thorough analysis, documentation, and management controls. It obviously does, just not to the same extent as safety critical systems.

3.2.2   The Need to Restrict Changes

Safety instrumented systems should be designed to allow very little, if any, human interaction. The operators had their chance to control the process with both the process control and alarm systems. (This topic will be discussed more thoroughly in Chapter 4.) If automatic process control systems and human intervention cannot make the required corrections, then the last line of defense should function automatically. About the only interaction that is allowed is maintenance overrides for performing work on portions of the system. People must know that there is a last line of defense, something that can be relied upon. These systems should have tightly controlled access. The last thing that should happen is that the system does not function because some lone operator disabled it without anyone else's knowledge. (It has happened.)

3.3   Separation of Control and Safety Systems

One of the most controversial topics in the industry is whether control and safety systems should be combined within the same system. In other words, should all safety functions be combined in the process control system? Proponents would argue that nowadays both systems are programmable, and process control systems are reliable and can be redundant, so why not? The answer is rather simple and does not hinge on "reliability." As we shall see, all of the standards, recommended practices, and guidelines in the industry recommend that *separate* systems be provided for process and safety control.

Trevor Kletz, one of the industry leaders in terms of process safety, has stated: "Safety systems such as emergency trips should be completely independent of the control system and, when practicable, hard-wired. If they are based on a computer, it should be independent of the control computer."[2] What some other documents state is discussed in the sections that follow.

3.3.1   Health and Safety Executive

The English couldn't make the message of separation any clearer. Their document, *Programmable Electronic Systems in Safety-Related Applications*, comes as a two-part volume. Part 1 is for the managers and is only 17 pages long. Figure 3-1, taken from it, takes an *entire page* in the document. (Part 2, the technical document for the engineers, is about 170 pages.) In other words, separate sensors, separate logic box, separate valves. "It is strongly recommended that separate control and protection systems are provided."[3]

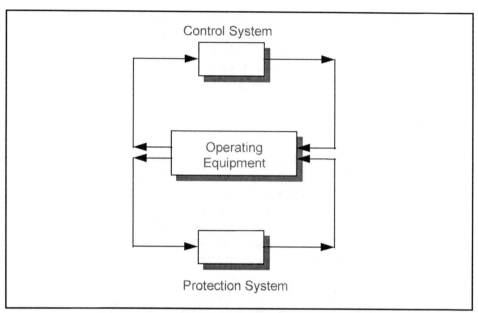

Figure 3-1   U.K. Health and Safety Executive Recommendations

3.3.2   American Institute of Chemical Engineers

The American Institute of Chemical Engineers' (AIChE) text is a set of "guidelines." About fifty pages of the text are devoted to what are termed *interlock systems* (along with some good background material). It contains statements that begin "Normally ..." that are always open to interpretation. Other such statements are very straightforward.

"Normally, the logic solver(s) are separated from similar components in the BPCS (Basic Process Control System). Furthermore, SIS input sensors and final control elements are generally separate from similar components in the BPCS." Moreover, "provide physical and functional separation and identification among the BPCS and SIS sensors, actuators, logic solvers, I/O modules, and chassis." [4]

3.3.3   International Electrotechnical Commission

The International Electrotechnical Commission (IEC) draft standard 61508 will be an international standard for all industries on the use of relay, solid-state, and programmable safety systems. It classifies four "Safety Integrity Levels" (SILs), with level 4 being the highest: "The separation of the safety related functions and the non safety related functions should be done whenever possible." [5] Separation is "highly recommended" for all four SILs.

The IEC document has gone through many changes, and portions are still in draft form. It has gone out for industry review twice. The industry has always favored the separation of process and safety control, but it is difficult to say exactly what the document will be like at the end of the review process.

3.3.4   ANSI/ISA

The ANSI/ISA standard S84.01-1996 states, "Sensors for SIS shall be separated from the sensors for the BPCS." Two exceptions are allowed: if there is sufficient redundancy, and if a hazard analysis determines that there are enough other safety layers to provide adequate protection. Also, "The logic solver shall be separated from the BPCS." There are exceptions for cases where it is not possible to separate control from safety functions, such as in certain rotating equipment.

Annex B of the standard contains the following statements: "It is generally necessary to provide separation between the BPCS and SIS functions … There are four areas where separation may be needed to meet the safety functionality and safety integrity requirements:

- logic solver
- field sensors
- final control elements
- communications with other equipment" [6]

3.3.5   American Petroleum Institute

The American Petroleum Institute's (API) recommended practice RP14C applies to the design of offshore platform shutdown systems. It prescribes how to determine the inputs and outputs for the shutdown system for different process units. It says nothing, however, about the design of the logic box (it assumes most logic systems are pneumatic): "The safety system should provide two levels of protection… The two levels of protection should be independent of and in addition to the control devices used in normal operation." [7]

It's surprising how open to interpretation some of these documents are. For example, would a relief valve and a gas detection system be considered "two levels of protection independent of and in addition to the control devices"? RP14C tells you how to determine input/output (I/O) and functional logic for the system. But it doesn't tell you what sort of logic box to use. People have installed pneumatic, relay, PLC, and TMR (triple modular redundant) systems on offshore platforms and have claimed that *all* have met RP14C. And, indeed, they may all have.

3.3.6   National Fire Protection Association

The National Fire Protection Association (NFPA) has written standards that apply to boiler and burner management systems. These might be viewed as examples of "prescriptive" standards. Prescriptive standards tend to have black-and-white statements that are easy to interpret and, therefore, comforting to some, such as "… shall not be combined with any other logic system." In the NFPA standard 8502, there is a very clear call for separation: "Requirement for Independence: The logic system performing the safety functions for burner management shall not be combined with any other logic system."[8]

3.3.7   Institute of Electric and Electronics Engineers

The Institute of Electric and Electronics Engineers (IEEE) has written standards that apply to the design of nuclear power generating stations. The process industry is not as restrictive as the nuclear industry, but the concept of separate safety layers is rather clear-cut here: "The safety system design shall be such that credible failures in and consequential actions by other systems shall not prevent the safety system from meeting the requirements."[9]

Some facilities use a quad redundant computer arrangement, dual in series with dual in parallel. And as if that weren't enough, some use *two* of these quad redundant systems together! The United States typically uses quad systems backed up with a "conventional analog" system.

An article in an English journal a few years back described the software testing process for the programmable safety system of an English nuclear power station. It took fifty engineers six months to proof test the system! It is doubtful that many companies in the petrochemical industry could afford twenty-five man-years to proof test each of their systems! Someone with the Nuclear Regulatory Commission (NRC) recently told one of the authors that it wasn't twenty-five man-years, rather more like several *hundred* man-years!

3.4   Common Cause Failures

Common cause failures are a single stressor that impacts multiple items or portions of a redundant system fail. One measure of such failures is referred to as a *beta factor*, which simply means the percentage of all identified failures in one "leg" of a system that might impact an entire redundant system. In other words, suppose a central processing unit (CPU) has a failure rate of 10E-6 failures/hour. If the CPU were triplicated and all three were placed side by side in the same chassis, certain failures might impact two or three CPUs at once. A beta factor of 10 percent would mean the system would act in a simplex manner with a failure rate of 1E-6 failures/hour.

Systematic failures (also called functional failures) are often viewed as another form of common cause failure. Systematic failures can be viewed as a single failure that impacts an entire system. Examples would be heat, vibration, design errors, maintenance errors, and the like. Although some common cause failures may be relatively easy to identify, the exact number or percentage of such failures is usually more difficult to quantify. How accurately can one predict, for example, how often an engineering design error might occur or that a maintenance technician might calibrate all three transmitters incorrectly?

One study in the nuclear industry found that 25 percent of all failures in nuclear power stations were related to common cause.[10] Investigations in redundant control systems have found common cause percentages between 10 percent and 30 percent. It is surprising that some highly redundant systems still have single points of failure. One networked computer system had seven-way redundant communications. Unfortunately, they all ran through the same fiber-optic cable, which when cut disabled the entire system.[11] Obviously, no one would *intentionally* design a system this way. The point is, things eventually grow to the point where one individual cannot "see" everything, and certain items inevitably fall into the cracks.

If control and safety functions are performed in the same system, there will always be the potential for common cause faults. The more the systems are physically separated, the more unlikely it will be that single failures can affect them both. A simple phrase sums up this thought up fairly well: "Don't put all your eggs in one basket." No matter how sturdy or reliable the basket may be, there will always be some unforeseen circumstances that will make you drop it.

Summary

Process control systems are active and dynamic, hence most faults are inherently self-revealing. Safety systems are passive or dormant, hence many faults are not self-revealing. Safety systems require either manual testing or effective self-diagnostics, something many general-purpose control systems do not incorporate.

In the distant past, process control and safety systems were implemented in separate, diverse technology systems. Pneumatic control systems are being replaced by software-based DCSs. Relay-based safety systems are being replaced by software-based PLCs. Incorporating both in one combined system offers the potential benefits of a single source of supply; simplified stores, maintenance, and training; and possibly lower cost. Standards, guidelines, and recommended practices from numerous industries, however, all strongly discourage this practice. The issue has little to do with the "reliability" of modern control systems.

If an organization chooses to diverge from the standards, it better have a valid, documented reason for doing so. A simple thought to keep in the back of your mind is "How would I justify this decision in court?" For if anything ever happened, that's what it might come down to. If your answer for differing from the standards was "It was cheaper that way," the court may not respond favorably.

References

1. ANSI/ISA, *Identification of Emergency Shutdown Systems That Are Critical to Maintaining Safety in Process Industries*, ANSI/ISA S91.01-1995 (Research Triangle Park, NC: ISA, 1995), ISBN 1-55617-570-1.

2. Trevor A. Kletz, *Computer Control and Human Error* (Houston, TX: Gulf Publishing, 1995), ISBN 0-88415-269-3.

3. U.K. Health and Safety Executive, *Programmable Electronic Systems in Safety-Related Applications, Part 1 - An Introductory Guide* (Sheffield, UK: U.K. HSE, 1987), ISBN 011-883913-6.

4. American Institute of Chemical Engineers, *Guidelines for Safe Automation of Chemical Processes* (New York: AIChE, Center for Chemical Process Safety, 1993), ISBN 0-8169-0554-1.

5. International Electrotechnical Commission, *Functional Safety: Safety-Related Systems*, draft standard 61508 (Geneva, Switzerland: IEC, 1997).

6. ANSI/ISA, *Application of Safety Instrumented Systems for the Process Industries*, ANSI/ISA S84.01-1996 (Research Triangle Park, NC: ISA, 1996), ISBN 1-55617-590-6.

7. American Petroleum Institute, "Recommended Practice for Analysis, Design, Installation, and Testing of Basic Surface Safety Systems for Offshore Production Platforms, API Recommended Practice 14C, fifth edition (Washington, DC: API, [March 1, 1994]).

8. National Fire Protection Association, *Prevention of Furnace Explosions/Implosions in Multiple Burner Boiler-Furnaces*, standard 8502 (Quincy, MA: NFPA, 1995).

9. Institute of Electrical and Electronics Engineers, *InterStandard Criteria for Safety Systems for Nuclear Power Generating Stations*, standard 603-1980 (Piscataway, NJ: IEEE, 1980).

10. David J. Smith, *Reliability, Maintainability and Risk (Practical Methods for Engineers)*, 4th ed. (London: Butterworth-Heinemann, 1993), ISBN 0-7506-0854-4.

11. P. G. Neumann, *Computer-Related Risks* (Reading, MA: Addison-Wesley, 1995), ISBN 0-201-55805-X.

4

PROTECTION LAYERS

That last trip generated over 17,000 alarm messages. The big guy wants your analysis and report in one hour.

Gruhn

Accidents rarely have a single cause. Accidents are usually a combination of rare events that people initially assumed were independent and would not happen at the same time. Take, for example, the worst chemical accident to date, Bhopal, India, where an estimated two thousand people died and two hundred thousand were injured.[1]

The material that leaked in Bhopal was MIC (methyl isocyanate). The release occurred from a storage tank, which held more material than was allowed by company safety requirements. Operating procedures specified using the refrigerant system of the storage tank to keep the temperature of the material below 5°C. A temperature alarm would sound at 11°C. The refrigeration unit was turned off, and the material was usually stored at nearly 20°C. The temperature alarm threshold was changed from 11°C to 20°C.

A worker was tasked to wash out some pipes and filters that were clogged. Water leaked past valves into a tank containing MIC. Temperature and pressure gauges that indicated abnormal conditions were ignored because they were believed to be inaccurate. A vent scrubber, which could have neutralized the release, was not kept operational because it was assumed to be unnecessary when production was suspended (as it was at the time). The vent scrubber was, in any event, inadequate to handle the size of the release. The flare tower, which could have burned off some of the material, was out of service for maintenance. It was also not designed to handle the size of the release. Material could have been vented to nearby tanks, but gauges erroneously showed them to be partially filled. A water curtain was available to neutralize a release, but the MIC was vented from a stack 108 feet above the ground, too high for the water curtain to reach.

Workers panicked and fled, ignoring four buses that were intended to evacuate employees and nearby residents. The MIC supervisor could not find his oxygen mask and broke his leg climbing over the boundary fence. The works manager, when informed of the accident, said in disbelief, "The gas leak just can't be from my plant. The plant is shut down. Our technology just can't go wrong. We just can't have leaks."

Investigations of industrial accidents have found that a large number have occurred during an interruption of production and while an operator was trying to maintain or restart production. In each case, the dangerous situation was created by a desire to save time and ease operations. And in each case, the company's safety rules were violated.[2]

The best and most redundant safety layers can be defeated by poor or conflicting management practices. Examples have been documented in the chemical industry.[1] One accident in a polymer processing plant occurred after operations bypassed all alarms and interlocks in order to increase production by 5 percent. In another, interlocks and alarms failed (at a normal rate), but this was not known because management had decided to eliminate regular maintenance checks of the safety instrumentation.

Figure 4-1 appears in a number of different formats in most all of the standards. It shows that there are various safety layers, some of which are prevention layers, others of which are mitigation layers. The basic concept is simple: "don't put all your eggs in one basket." Some refer to this as "defense in depth."

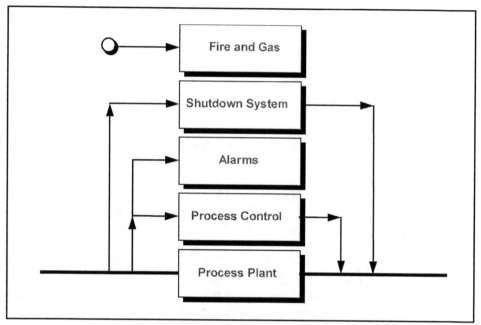

Figure 4-1   Protection Layers

Safety instrumented systems (SIS) are designed to monitor the process and control outputs in order to prevent or mitigate hazardous events. It is worth drawing the distinction between the prevention layers (which are designed to prevent the hazard from occurring in the first place) and mitigation layers (which are designed to contain or lessen the consequences of the hazard once it has happened).

4.1   Prevention Layers

The discussions of prevention and mitigation layers in sections 4.1 and 4.2 are examples only. The options discussed are not intended to be viewed as all the possible layers that may be implemented or should be used in any one facility.

4.1.1   Process Plant Design

The process plant itself must be designed with safety in mind. This is why HAZOP (hazard and operability studies) and other reviews, such as fault trees, checklists, what-ifs, and so on, are performed.

A major thrust within the process industry is to design inherently safe plants. Don't design a dangerous plant with the intention of throwing on lots of "Band-Aids" to "fix" the problem. Design it so the Band-Aids aren't even necessary. Work with low-pressure designs, low inventories, nonhazardous materials, and the like.

It's surprising how much resistance there is to this and most other sorts of change. The chief cry is "But we can't *afford* to do all that!" Information one of the authors has seen in various industries indicates that inherently safer systems *may* have a higher initial cost (although not always), but they offer a *lower* cost of ownership *over the life* of any project. The same applies to quality management. Think about how many companies said they couldn't afford that either. Now, you can't afford *not* to have it.

Eliminating or reducing hazards often results in a *simpler* design, which may, in itself, reduce risk. The alternative is to add protective equipment to control hazards, which usually adds complexity.

One example of the "rebellion" against such designs can be found in everyday life. The Refrigerator Safety Act was passed because children were suffocating after becoming trapped while playing in unused refrigerators. Manufacturers insisted they could not afford to design safer latches. When forced to do so, they introduced simple, magnetic latches. These permitted the door to be opened from the inside, thus *eliminating* the hazard, and they were also *cheaper* than the older type.[2]

4.1.2   Process Control System

The process control system is the next layer of safety. It controls the plant for optimum fuel usage, product quality, and the like and keeps all variables (e.g., pressure, temperature, level, flow, etc.) within safe bounds. Some are reluctant to consider the process control system as a "safety layer," but the authors of this book have no such problem, so long as it is not the *only* safety layer.

Automation usually does not eliminate humans from the system and frequently raises their tasks to new levels of complexity. If computers are being used to make more and more decisions because human judgment and intuition are not satisfactory, then it may be an error to have a human act as the final arbiter. Experience has shown that humans make poor monitors of automated systems. Tasks that require little active operator action may result in lowered alertness and vigilance and can lead to complacency and an overreliance on automated systems. Long periods of passive monitoring can cause operators to be unprepared to act in emergencies. Some have commented that "computer control turns operators into morons." One way to solve this problem would be to involve operators in safety analysis and design decisions up front and throughout development–that is, involve them more, not less.

4.1.3 Alarm Systems

If the process control system fails to perform its function (for any number of reasons) alarms may be used to alert the operators that some form of intervention is required on their part.

Alarm and monitoring systems should (1) detect problems as soon as possible, at a level low enough to ensure that action can be taken before hazardous states are reached; (2) be independent of the devices they are monitoring (they should not fail if the system they are monitoring fails); (3) add as little complexity as possible, and (4) be easy to maintain, check, and calibrate.

Alarm and monitoring systems are the safety layer at which people get actively involved. Operators will usually be required for the simple reason that not *everything* can be automated. It is essentially impossible for designers to anticipate *every* possible set of conditions that might occur. Human operators may need to be considered since only they will be flexible and adaptable enough in certain situations.

This is a two-edged sword, however, because events not considered in the design stage will no doubt also not be included in operator training either. On the other hand, simply following procedures blindly has resulted in accidents. Deviating from the rules when necessary is a hallmark of experienced people, but it is bound to lead to occasional human error and related blame after the fact.

4.1.3.1 Human Reliability

One of the authors of this book has heard some people say they did not want automated systems in their plants controlling safety. They wanted to rely on people who were educated and trained in the operation and dynamics of their process. Although this may be acceptable for normal routine operations (some might argue this), it is *not* the recommended scheme for critical emergency situations.

For example, accidents have occurred because

1. Operators did not believe rare events were real or genuine, or
2. Operators were overloaded with information and failed to act.

People have been, and will continue to be, directly responsible for some accidents. Some in the industry, most notably Trevor Kletz, have done an excellent job documenting such cases.[1] Hopefully, the rest of the industry will learn from these examples and not repeat them.

For example, there have been cases where the operators saw the alarm, knew what it meant, and *still* took no action. Either the alarm was considered a "nuisance" alarm ("Oh, we see that all the time"—this was one of many problems at Bhopal) or they waited to see if anything else would happen (sometimes with catastrophic results).

When things do go wrong, they tend to cascade and escalate. One of the authors knows of one plant where there was a shutdown, and the DCS printed out *seventeen thousand* alarm messages! Overwhelming the operators with this much information is obviously detrimental. *Too* much information is *not* a good thing.

When faced with life-threatening situations that require that decisions be made within a minute, people tend to make the wrong decision 99 percent of the time. In other words, during emergencies people are about the *worst* thing to rely on, no matter how well they have been trained.

4.1.4   Shutdown/Interlock/Instrumented Systems

If the control system and the operators fail to act, automatic shutdown systems take control. These systems are usually completely separate, with their own sensors and valves. (We discussed the separation issue in Chapter 3.) These systems require a higher degree of security so as to prevent inadvertent changes and tampering and a greater level of fault diagnostics. The focus of this book is on these systems.

4.2   Mitigation Layers

Prevention layers are designed to prevent the hazard form occurring in the first place. Mitigation layers are designed to contain or lessen the consequences of the hazard once it has already happened.

4.2.1   Fire & Gas Systems

If the safety instrumented system fails and an accident ensues, fire & gas systems may be used to mitigate or lessen the consequences of the event. In the United States, these are traditionally "alarm-only" systems–they do not take any automatic control actions. Typically, the fire crews must go out and manually put out the fire. Outside the United States, these systems frequently take some form of control action or may be integrated with the shutdown system.

One major difference between shutdown and fire & gas systems is that shutdown systems are usually normally energized (and deenergize to shutdown), while fire & gas systems are normally deenergized (and energize to take action). The reasoning for this is actually rather simple. Shutdown systems are designed to bring the plant to a safe state, which usually means stopping production. Nuisance trips (shutting the plant down when nothing is actually wrong) is economically detrimental because of the lost production during downtime but is generally not catastrophic in terms of safety. Actually, studies have shown that while shutdown/start-up operations only account for about 4 percent of the total operating time, about 25 percent of all accidents happen during that 4 percent.[3] Fire & gas systems are designed to protect equipment and people. The spurious operation of these systems can destroy certain pieces of equipment and possibly even result in deaths (e.g., an unannounced halon or CO_2 dump in a control room). If the systems are normally deenergized, such failures become highly unlikely.

4.2.2   Containment Systems

There are other mitigation layers not shown in Figure 4-1. For example, if an atmospheric storage tank were to burst, dikes could be available to contain the release. In nuclear power plants, reactors are usually housed in containment buildings to prevent accidental releases. (The Soviet reactor at Chernobyl did not have a containment building, whereas the U.S. reactor at Three Mile Island did.)

4.2.3   Evacuation Procedures

In the event of a catastrophic release, evacuation procedures are used to evacuate plant personnel from the area and, if necessary, even the outside community. Although these are procedures only and not a physical system (apart from the sirens), they may still be considered one of the overall safety "layers."

4.3   Diversification

Investors understand the need for diversification. If investors invest their entire life savings in one stock account and that account goes bust, they are out their entire savings. It is safer to spread the risk among multiple stock accounts; that way, if any one fails, the others will hopefully protect against total loss. The same applies to safety layers. Again, don't put all your eggs in one basket, no matter how good the basket may be. Everything fails, it's just a matter of when. The more layers there are the better. In addition, each layer should be as simple as possible, and the failure of one layer should not prevent another layer from performing its intended function.

Summary

Accidents are usually a combination of rare events that people initially assumed were independent and would not happen at the same time. One method of protecting against such instances is to implement multiple, diverse safety layers, making it harder for any one event to precipitate a hazardous condition.

One should implement multiple, independent safety layers. Some layers are in place to prevent accidents from occurring (prevention layers); others are in place to lessen the consequences of an event once it happens (mitigation layers).

The more safety layers the better, although the best defense is to remove the hazard in the initial design phase. Inherently safe designs can result in simpler designs with lower overall costs.

References

1. Trevor A. Kletz, *Lessons from Disaster: How Organizations Have No Memory and Accidents Recur* (Houston, TX: Gulf Publishing, 1994), ISBN 0-88415-154-9.

2. Nancy G. Leveson, *Safeware: System Safety and Computers* (Reading, MA: Addison-Wesley, 1995), ISBN 0-201-11972-2.

3. *Oil and Gas Journal*, August 27, 1990.

4. American Institute of Chemical Engineers, *Guidelines for Safe Automation of Chemical Processes* (New York: AIChE, Center for Chemical Process Safety, 1993), ISBN 0-8169-0554-1.

DEVELOPING THE SAFETY
REQUIREMENT SPECIFICATIONS

Nothing can go wrong (click), go wrong (click), go wrong (click) ...

5.1 Introduction

Once the need for a safety instrumented system (SIS) is identified and the target safety integrity level (SIL) has been established for each safety function, the safety requirement specifications (SRS) have to be prepared for the system. The SRS consists of two parts:

1. the functional requirement specifications
2. the integrity requirement specifications

The functional requirements describe the process interlocks; for example, low pressure in vessel A will automatically close valve B. It defines *what* the safety system does. The integrity requirements describe the probability or likelihood that valve B will close when the pressure in vessel A is low. It defines the *capability* or *how well* the safety system works. When the word *specification* is used in this chapter, it refers to both the functional requirement and the integrity requirement specifications.

It is best to design plants that are inherently safe. In other words, design the plant so that hazards won't exist. Unfortunately, this is not always possible. There are times when the only way to prevent a hazard is to add instrumentation to monitor the process conditions and to operate devices to take proper action. Deciding what process parameters are to be monitored, what the safety control actions are to be, and how well the system will operate are all reflected in the functional and integrity specifications.

In Chapter 1 we noted that the English Health and Safety Executive examined over thirty accidents that were the direct result of control and safety system failures in a variety of different industries.[1] The findings, which are summarized in Figure 1-1, show that 44 percent of accidents are caused by incorrect specifications (both the functional specification as well as the integrity specification). The second highest percentage of failures (20 percent) is the result of changes after commissioning. The importance of ensuring that the specifications of the safety system are correct cannot be therefore overemphasized.

This chapter identifies some of the problems associated with developing specifications, reviews ANSI/ISA S84.01-1996 requirements, and includes typical documentation that can be used to document the specification requirements.

5.2   Why Are 44 Percent of Accidents Caused by Incorrect Specifications?

To answer *why* incorrect specifications were the most common cause of accidents one has to look not only at the specific reasons but also at the systems and procedures in place to produce the specification. Following the design life cycle model outlined in Chapter 1, the following activities must occur before the specification is prepared:

1. Conceptual process design
2. Hazard analysis and risk assessment
3. Application of non-safety-instrumented-system layers
4. SIS classification requirements if a SIS required

These steps, 1 through 4, lay the foundation for a correct specification.

The systems and procedures that exist within an organization to complete these four steps must be adequate. Irrespective of how much effort is spent in ensuring that the specifications reflect the requirements, the fundamental design basis for the preparation of the specification will be incorrect if the systems and procedures within an organization are nonexistent or inadequate.

The purpose of this book is not to define process design procedures or hazard analysis techniques. Numerous publications on these subjects are available from the Center for Chemical Process Safety of the American Institute of Chemical Engineers and from other sources on process design and hazard analysis. It should be pointed out that identifying and assessing hazards is not what makes a system safe. It is the information obtained from the hazard analysis and the actions subsequently taken that determine system safety. SIL classification methods are covered in chapter 6.

In addition to the systemic and procedural issues just mentioned, the following are some more specific reasons why such a large percentage of failures are attributed to incorrect specifications:

- management systems
- procedures
- scheduling of assessment
- participation of key personnel in review process
- responsibilities not well defined
- training and tools
- complexity and unrealistic expectations
- incomplete documentation
- inadequate final review of specification
- unauthorized deviation from specification

The details of these mentioned are described the following sections.

5.2.1   Management Systems

A management system, which is basically a system for ensuring effectiveness, must be in place for the life cycle model to be effective. A typical management system usually consists of the following steps:

1. identifying the activities in the life cycle model, and the expected results of those activities

2. setting the performance standards by which the results will be compared

3. measuring the actual performance

4. evaluating and comparing actual performance versus set standards

5. making commendations or corrections based on evaluation

It is sometimes easy to develop standards, procedures, and the like but difficult to develop and implement a management system to ensure full compliance. The root cause of most problems can usually be traced back to nonexistent or ineffective management systems.

F. Bird and G. Germain's *Practical Loss Control Leadership* is an excellent text on dealing with management systems.[2] Note: The term *management system*, although widely used, is a misnomer. These systems in many cases have little to do with "management." A better term is *managing systems*.

5.2.2   Procedures

For all activities, procedures have to be developed, reviewed, approved, and understood in order to effectively complete the hazard assessment and SIL classification. Are the procedures intended to cover safety only or are they intended to cover other protective systems, for example, equipment, environment, or quality? This should be clearly reflected in the procedures.

Hazard assessment usually involves techniques such as fault tree analysis (FTA), failure mode and effects analysis (FMEA), hazard and operability studies (HAZOP), and other similar processes. The HAZOP technique is widely used, and it is easy to develop procedures to satisfy any organization's requirements.

SIL classification techniques are not as well established, but large organizations usually have the resources to develop the techniques and procedures. Smaller organizations may have to rely on outside consultants for assistance.

5.2.3   Scheduling of Assessment

In the rush to have the safety assessment completed this activity is sometimes carried out without the process design requirements being well defined or understood. We have to ensure that the process design and the documentation required for the assessment have been completed to such an extent that a meaningful safety analysis could be done.

The assessment must not be a mere formality to satisfy the project management schedule and guidelines. The SIL classification should be an extension of hazard review so that the synergies and understanding developed during the safety assessment are extended to the classification.

5.2.4 Participation of Key Personnel in Review Process

Most systems in process plants are far too complex for any one person to be a total expert on them. One person cannot be a master of the chemistry of a process, the materials used to contain it, the instrumentation used to monitor it, the control systems used to control it, the rotating equipment used to drive it, the electrical systems used to power it, or the safety techniques used to analyze it.

For this very reason, HAZOP and other hazard analysis techniques require a multidisciplinary team. Persons representing the project, process, operations, safety, maintenance, instrumentation, and electrical groups are typically required. These studies analyze what might go wrong in the operation of a plant. If a hazard is identified and considered significant, a decision must be reached about how to prevent or mitigate the hazard.

As we proceed through the different phases of the hazard analysis and SIL determination, we have to be careful that each phase is not regarded as a distinct activity and upon completion of that particular phase that the results are not passed to another team over a "solid brick wall."

It is essential that a team approach be developed for the reviews.

5.2.5 Responsibilities Not Well Defined

At the beginning of section 5.2 we listed the four activities that have to be completed before preparing the specification. For each activity, the individual with the lead/responsible role must be identified and agreed upon. Typical lead roles for each of the four activities are as follows:

Activity	Lead Role Responsibility
Conceptual process design	Process design
Hazard analysis and risk assessment	Safety engineering
Application of non-SIL layers	Process design
SIL classification requirements if SIS required	Safety engineering

The individuals who have the lead role responsibility must clearly understand their role and ensure that the analysis is completed in accordance with the established procedures.

5.2.6   Training and Tools

The team involved with the assessment has to be adequately trained and provided with the required tools. For example, before the assessment, control and electrical personnel may need to receive the required process training or process personnel may need to be adequately trained on the type and capabilities of the control systems. This is essential for ensuring the proper dialog and understanding during the review process. Before the analysis, the capabilities and special training needs of each team member must be identified.

5.2.7   Complexity and Unrealistic Expectations

Generally speaking, the simpler a design the better. Complex designs are more difficult to understand, visualize, and review. Take as an example one accident described by Trevor Kletz.[3]

An accident in a process plant involved a highly hazardous ethylene oxide process. Three valves in series were installed, with two bleed valves between them. As if that weren't enough, these five valves were duplicated so the system could be tested while the plant was on line without interrupting production.

The operation of the system was carried out automatically by a computer. As a result of a software error, the computer momentarily operated the valves incorrectly. The ensuing explosion and fire involving three tons of gas caused extensive damage and resulted in lost production for several months. The software error could be attributed to complexity.

If a safety system is required to mitigate a hazard, only the simplest requirements should be stated. The team personnel should refrain from establishing solutions that are too complex. This leads to unrealistic and unwarranted expectations.

5.2.8   Incomplete Documentation

Process control/instrumentation personnel are usually responsible for preparing the safety requirement specifications based on the safety assessment.

Item	Details of Requirement
Documentation and Input Requirements	
P&IDs	
Cause-and-effect diagram	
Logic diagrams	
Process data sheets	
Process information (incident cause, dynamics, final elements, etc.) of each potential hazardous event that requires an SIS	
Process common cause failure considerations such as corrosion, plugging, coating, etc.	
Regulatory requirements impacting the SIS	
Other	
Functional Requirements	
The definition of the safe state of the process, for each of the identified events	
The process inputs to the SIS and their trip points	
The normal operating range of the process variables and their operating limits	
The process outputs from the SIS and their actions	
The functional relationship between process inputs and outputs, including logic, math functions, and any required permissives	
Selection of deenergized to trip or energized to trip	
Consideration for manual shutdown	
Action(s) to be taken on the loss of energy source(s) to the SIS	
Response time requirement for the SIS to bring the process to a safe state	
Response action to any overt fault	
Human-machine interfaces requirements	
Reset function(s)	
Other	
Safety Integrity Requirements	
A list of the safety function(s) required and the SIL of each safety function	
Requirements for diagnostics to achieve the required SIL	
Requirements for maintenance and testing to achieve the required SIL	
Reliability requirements if spurious trips may be hazardous	
Failure mode of each control valve	
Failure mode of all sensors and transmitters	
Other	

Table 5-1   Summary Information That Should Be Included in Safety Requirement Specifications

The method for documenting the safety requirement specifications has to be simple enough for all individuals to be able to understand what needs to be provided. Table 5-1 summarizes the information that should be included in the specification.

One very simple but effective documentation tool is the cause-and-effect diagram (see Table 5-2). This diagram clearly shows the input/output relationship for the safety system, the trip set points, the SIL for each function, and any special design requirements.

TAG #	Description	SIL	INSTRUMENT RANGE	TRIP VALUE	UNITS	OPENS VALVE FV-1004	OPEN VALVE XV-1005	STOP PUMP P-1007	CLOSE VALVE PV-1006	NOTES
FL-1000	Flow to main reactor R-100	2	0-200	100.0	gpm	X		X		1
PH-1002	Reactor internal pressure	2	0-800	600	psi	X	X			1, 2
XS-1003	Loss of control power					X	X	X	X	
XS-1004	Loss of instrument air					X	X	X	X	
HA-1005	Manual shutdown					X	X	X	X	

Notes:
1. Two-second delay required before closing FV-1004.
2. Reset located at valve PV-1006.

Table 5-2　　　Cause-and-Effect Diagram

The specification should state *what* is to be achieved, not necessarily *how* to achieve it. This allows others the freedom to decide how best to accomplish the goals and also allows for the development and application of new knowledge and techniques.

5.2.9　Inadequate Final Review of Specification

Upon completion of the specification it should be reviewed and approved by all parties involved with the SIL classification to ensure that they understand it and can give it final approval. We should not assume that everyone has the same understanding at the SIL review. This final review will give everyone an opportunity to reflect on the decisions and to clarify any misunderstandings.

5.2.10   Unauthorized Deviation from Specification

Once the specification is approved, no deviations should be made unless a formal change procedure is followed (see Chapter 13, "Managing Changes to a System"). It is sometimes tempting to make unauthorized changes to avoid cost or schedule hurdles.

The specification must be updated when changes are made during the course of the project. No changes to, or deviation from, the specification must be made without approval from the project team.

5.3   ANSI/ISA S84.01-1996 Requirements

ANSI/ISA S84.01-1996 provides an excellent summary of the specification requirements.[4] This section shows the requirements of the standard (shown in italics).

Input Requirements

The information required from the Process Hazards Analysis (PHA) or process design team to develop the Safety Requirement Specifications includes the following:

- *Process information (incident cause, dynamics, final elements, etc.) of each potential hazardous event that requires an SIS.*

- *A list of the safety function(s) required and the SIL of each safety function.*

- *Process common cause failure considerations such as corrosion, plugging, coating, etc.*

- *Regulatory requirements impacting the SIS.*

Safety engineering personnel normally supply all this data, except the SIL for each function. The SIL for each safety function is usually determined as part of the SIL classification study and not as part of the hazard assessment, but, as stated in section 5.2.3, it is preferable that the two activities be merged.

Safety Functional Requirements

The safety functional requirements shall include the following:

- *The definition of the safe state of the process, for each of the identified events.*

- *The process inputs to the SIS and their trip points.*

- *The normal operating range of the process variables and their operating limits.*

- *The process outputs from the SIS and their actions.*

- *The functional relationship between process inputs and outputs, including logic, math functions, and any required permissives.*

- *Selection of deenergized to trip or energized to trip.*

- *Consideration for manual shutdown.*

- *Action(s) to be taken on loss of energy source(s) to the SIS.*

- *Response time requirement for the SIS to bring the process to a safe state.*

- *Response action to any overt fault.*

- *Human-machine interfaces requirements.*

- *Reset function(s).*

Safety Integrity Requirements

Safety integrity requirements shall include the following:

- *The required SIL for each safety function.*

- *Requirements for diagnostics to achieve the required SIL.*

- *Requirements for maintenance and testing to achieve the required SIL.*

- *Reliability requirements if spurious trips may be hazardous.*

These requirements are usually obtained from the process hazard analysis (PHA) and the SIL classification.

5.4 Documenting the Specification Requirements

Since the specification will provide the guidelines for preparing the design requirements, all the required information should be included as a complete package. The following four items should be considered the key documents to be included in the package.

1. The **process description**. This should include the following:

   - P&IDs (process and instrumentation diagrams)

   - description of the process operation

   - process control description including control system design philosophy, type of controls, operator's interface, alarm management, and historical data logging

- special safety regulations--should include the corporate, local, state/provincial, or federal requirements

- reliability, quality, or environmental issues

- list of operational or maintenance issues

2. **Cause-and-effect diagram:** This can be used to document the functional and the integrity requirements. These diagrams are very simple documents that are easily understood by all disciplines. See Table 5-2 for a typical diagram.

3. **Logic diagrams:** Logic diagrams can be used to supplement the cause-and-effect diagrams for complex and time-based functions and for complex sequences that cannot be easily described in the cause-and-effect diagram. Logic diagrams could be produced in conformance with ANSI/ISA standard S5.2.

4. **Process data sheets:** Process data sheets ensure that the process information required by process control engineering to complete the instrument specification sheets is well documented.

5.4.1   Safety Requirement Specifications–Summary Table

Table 5-1 can be used to summarize the safety requirement specifications. The "Details of Requirement" column identifies whether information pertaining to each item should be provided for this application. If it is required, any special comments should also be included. Other related issues are summarized in the design checklist (see Chapter 15).

Summary

The safety requirement specifications (SRS) consist of two parts: the functional requirement specifications and the integrity requirement specifications. The functional requirements describe the process interlocks, and the integrity requirements describe the probability of the safety system working.

The English Health and Safety Executive examined over thirty accidents that were the direct result of control and safety system failures in a variety of different industries, and their findings showed that 44 percent of accidents were caused by incorrect specifications (both the functional specification as well as the integrity specification).

The importance of ensuring that the specifications of the safety system are correct cannot therefore be overemphasized.

References

1. U.K. Health and Safety Executive, *Programmable Electronic Systems in Safety-Related Applications, Part 1: An Introductory Guide* (Sheffield, UK: U.K. HSE, 1987), ISBN 011-883913-6.

2. F. E. Bird Jr. and George L. Germain, *Practical Loss Control Leadership* (Loganville, GA: International Loss Control Institute), ISBN 0-88061-054-9.

3. Trevor A. Kletz, *Computer Control and Human Error* (Houston, TX: Gulf Publishing, 1995), ISBN 0-88415-269-3.

4. ANSI/ISA, *Application of Safety Instrumented Systems for the Process Industries*, ANSI/ISA S84.01-1996 (Research Triangle Park, NC: ISA, 1996), ISBN 1-55617-590-6.

6

DETERMINING THE SAFETY INTEGRITY LEVEL

"The man who insists upon seeing with perfect clearness before deciding, never decides."

— *H. F. Amiel*

Many industries have the need to evaluate and rank risk so management decisions may then be made regarding various design options. For example, how remote (if at all) should a nuclear facility be from a large population zone? What level of redundancy is appropriate for a military aircraft's nuclear weapons control system? How strong should jet engine turbine blades be to protect them from flying birds? Based upon known failure rate data, how long should a warranty period be? Ideally, decisions such as these would be made based upon mathematical analysis. Realistically, the quantification of *all* factors is extremely difficult, if not impossible, and subjective judgment and experience may still be considered.

Military organizations were some of the first groups to face such problems. For example, when someone has to press the button launching a nuclear missile, possibly ending civilization as we know it, the probability that the electronic circuitry will work properly should be rated as something other than just "high." The U.S. military developed a standard method for categorizing risk—*Standard Practice for System Safety Requirements* (MIL-STD 882)—that has been adapted by other organizations and industries in a variety of formats.[1]

6.1   Who's Responsible?

Since determining the safety integrity level (SIL) is mentioned in the various standards, many assume that task therefore falls within the responsibility of the control system engineer. This is *not* the case. Evaluating the process risk and determining the appropriate integrity level is a responsibility of a multidisciplinary *team*, not any one individual. A control system engineer may (and should) be involved, but the review process will also require other specialists, such as those typically involved in a HAZOP review. In fact, many believe determining the SIL should be done as part of a HAZOP.

6.2   A Caution

In this area, there are no clearly defined issues or answers that could be categorized as either "right" versus "wrong" or "black" versus "white." There are many ways to evaluate process risk, none more correct than another. Various industry documents describe more than one qualitative method for evaluating risk and determining safety instrumented system (SIS) performance.[2] Several of the documents are reluctant to describe quantitative methods (for both ethical and legal reasons), but they are just as valid.

6.3   Common Issues

No matter what risk analysis method is chosen, several factors are common to all. For example, all categorization methods involve evaluating the two components of risk (probability and severity), usually by splitting them into different levels.

There are different hazards associated with each process unit, and each hazard has its associated risk (probability and severity). Take, for example, a vessel where one is measuring pressure, temperature, level, and flow. The pressure measurement is probably meant to prevent an overpressure condition and explosion. This would have a corresponding level of risk (probability and severity). Low flow might only result in a pump burning up. This would have a completely different severity rating and therefore a different SIL requirement. High temperature might result in an off-spec product. This would also have a completely different severity rating and a different SIL requirement. What this means is that one should *not* try to determine the SIL for an *entire* process unit but should rather determine the SIL for *each safety function*.

6.4   Method No. 1 (Qualitative)

The first method described here originated with the U.S. military (MIL-STD 882). It categorizes frequency and severity using five different qualitative levels. Five levels is merely an example, for one can easily choose fewer or more levels.

6.4.1   Evaluating the Frequency

The frequency or probability of an event may be ranked from low to high, from improbable to frequent, or by whatever measure may be appropriate. As shown in Table 6-1, this ranking may be for a single item or for a group of items, for a single process unit or for an entire plant.

Level	Descriptive Word	Qualitative Frequency	Quantitative Frequency
5	Frequent	A failure that can reasonably be expected to occur more than once within the expected lifetime of the plant.	Freq > 1/10 per year
4	Probable	A failure that can reasonably be expected to occur within the expected lifetime of the plant.	1/100 < Freq < 1/10 per year
3	Occasional	A failure with a low probability of occurring within the expected lifetime of the plant.	1/1,000 < Freq < 1/100 per year
2	Remote	A series of failures with a low probability of occurring within the expected lifetime of the plant.	1/10,000 < Freq < 1/1,000 per year
1	Improbable	A series of failures with a very low probability of occurring within the expected lifetime of the plant.	Freq < 1/10,000 per year

Table 6-1   Risk Frequency (Example Only)

6.4.2   Evaluating the Severity

Severity may also be categorized according to the different factors at risk: people, capital equipment, production, and the like. The numbers shown in Tables 6-1 and 6-2 are *examples only* and are *not* intended to represent, or even imply, any kind of recommendation.

Level	Descriptive Word	Potential Severity/Consequences		
		Personnel	Environment	Production/Equipment
V	Catastrophic	Death outside plant	Detrimental off-site release	Loss > $1.5M
IV	Severe	Death in plant	Nondetrimental off-site release	Loss between $1.5M and $500K
III	Serious	Lost time accident	Release on site - not immediately contained	Loss between $500K and $100K
II	Minor	Medical treatment	Release on site - immediately contained	Loss between $100K and $2,500
I	Negligible	First aid treatment	No release	Loss < $2,500

Table 6-2   Risk Severity (Example Only)

6.4.3   Evaluating the Overall Risk

The two sets of frequency and severity numbers may then be combined into an *x-y* plot as shown in Table 6-3. The lower left corner represents "low" risk (low probability and low frequency); the upper right corner represents "high" risk (high probability and high frequency). Note that the differentiating borders can be rather subjective.

High Risk

Severity	Frequency				
	1	2	3	4	5
V	1-V	2-V	3-V	4-V	5-V
IV	1-IV	2-IV	3-IV	4-IV	5-IV
III	1-III	2-III	3-III	4-III	5-III
II	1-II	2-II	3-II	4-II	5-II
I	1-I	2-I	3-I	4-I	5-I

Table 6-3   Overall Risk (Example Only)

Low Risk   Medium Risk

6.4.4   Evaluating the Effectiveness of Additional Layers

A third axis appears in both the American Institute of Chemical Engineers' (AIChE) Center for Chemical Process Safety (CCPS) text and the ANSI/ISA S84.01-1996 standard that does not appear in the military standard (see Figure 6-1). This third axis is designed to take account of additional safety layers. For example, does the system under consideration have any form of additional layers of protection beyond it? In other words, if the system under consideration fails, will anything else be able to prevent the hazard? If not, then you are restricted to the first level. However, if there are additional layers, perhaps you can lower the design requirements of the system under study. Instead of requiring level 3 performance, perhaps the requirements could be lowered to those for level 2. Again, the borders are rather subjective.

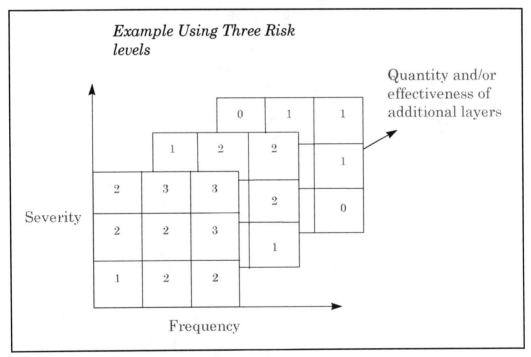

Figure 6-1   Impact of Additional Safety Layers

Many are concerned with the qualitative judgments involved, for every company can take the liberty of defining the levels and rankings a bit differently. OSHA representatives have admitted they are not process experts–they just want to see the paperwork trail your company has left behind and the decision process showing how you got there. The fact that different locations within the same company can define all of this differently does not really matter (at least to the standards bodies), even though it is naturally a cause for concern. The industry simply has not reached (and may never reach) consensus on all of the subjective decisions and rankings involved.

There is a simple, yet rather crass thought to keep in the back of your mind during this whole process–"How will we defend this decision in a court of law?" As scary as that thought may seem, it sums up the bottom line rather well. Think about it. If there is an accident and people are injured, the case will go to court. Someone somewhere will need to be able to defend their design decisions. If asked why a certain decision was made and the response is "Well, we weren't sure what to do, but that's what our vendor recommended," the court's response may naturally be less than forgiving.

6.4.5 Combining Method No. 1 Factors to Determine the Safety Integrity Level

The AIChE CCPS text and ANSI/ISA S84.01-1996 standard have categorized three overall safety integrity levels: essentially, low, medium, and high. As stated earlier, the levels are somewhat subjective and may be defined differently by different companies. The next step in the process is to relate the level of risk to the level of performance of the safety system. The CCPS, ISA, and International Electrotechnical Commission (IEC) documents all do this using a table, such as Table 6-4. The basis for the performance goals listed in Table 6-4 has been a subject of controversy. In other words, how was Table 6-4 "calibrated"? Perhaps the numbers should be shifted up a row or perhaps down a row.

Risk Level	IEC/ISA/AIChE Safety Integrity Level	Required Safety Availability	Probability of Failure on Demand	Risk Reduction Factor
(Note 1)	4	> 99.99%	< .0001	> 10,000
High	3	99.9 - 99.99%	.001 - .0001	1,000 - 10,000
Medium	2	99 - 99.9%	.01 - .001	100 - 1,000
Low	1	90 - 99%	.1 - .01	10 - 100

Notes:
1. SIL 4 is not used in the process industry documents. It is intended to represent other industries, such as transportation, aerospace or nuclear.
2. Field devices are *included* in the above performance requirements.

Table 6-4   Correlation between Overall Risk Level and Required Safety System Performance

6.5   Method No. 2 (Qualitative)

The IEC has adopted a slightly different method for evaluating risk based upon a German national standard (DIN/VDE 19250) (see Figure 6-2). One can see from Figure 6-2 that the wording is purposely vague and open to interpretation.

Figure 6-2 was intended for ranking the risk to personnel, although one could develop similar decision trees for other sorts of risk. One starts at the left end of the chart. For the case in question, what are the consequences to the people involved? Next, what is their frequency and exposure to the particular risk? Keep in mind that even though a process might be continuous, the personnel might only be in the particular area ten minutes out of the day, so their *exposure* would *not* be continuous. Next, one considers the possibility of avoiding the accident. In other words, is the process reaction time slow enough for the people to take action? Do the operators have local indications so they know what is going on in the process? Are they adequately trained to know what to do when an abnormal situation develops? Are there evacuation routes so they can leave the area in case of the hazard? Then one considers the probability of the event actually occurring. This primarily has to do with the additional safety layers. In other words, considering all of the other independent safety layers, what is the actual likelihood of this particular event happening? One can see that all of this is rather subjective.

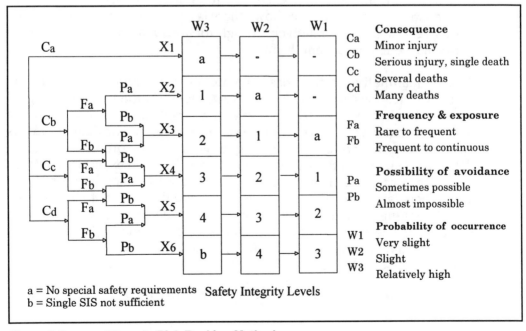

Figure 6-2   Alternate Risk Ranking Method

One may also question the "calibration" of Figure 6-2. Who is to say this chart should also not be shifted up or down one row? No doubt the Germans had some sort of rationale for developing it this way.

The system performance requirements for each SIL are the same as those outlined in Table 6-4.

6.6   Method No. 3 (Quantitative)

Some engineers are not comfortable with the qualitative methods for determining SIL outlined in the previous sections. The repeatability of the methods is questionable. For example, if you give the same problem to five different design groups and two groups come up with SIL 1, two groups with SIL 2, and one group with SIL 3, what might one conclude? Such differences would result in vastly different SIS designs. Are some of the groups "right" and others "wrong"? If five different aircraft design groups were all given the same design requirements (e.g., carry 250 people six thousand miles nonstop), they would certainly come up with different design concepts, but one would like to think there would be some similarities. In actuality, determining the SIL of a process is more akin to determining the aircraft design requirements rather than the actual implementation of that design.

Some are more comfortable with a fully quantitative method. For example, if your company does not want the hazard to happen more often than once in a thousand years (see Chapter 2 for a more complete discussion of the statistics), and experience shows there is a process demand on average once a year, then it is a relatively straightforward matter to determine the performance required from the SIS.

A fully quantitative method has been around for decades. It matches the required safety system performance based upon (1) the target safety goal and (2) the demand rate. An example of a target safety goal would be a mean time between accidents of three thousand years. The demand rate is an estimate of how often something goes wrong in the process; how often there is a demand on the safety system to function. An example would be a compressor that, on average, overpressures a vessel once per year. If you know there is a demand once per year but you don't want to tolerate the vessel bursting more than once in three thousand years (on average), then it is easy to specify the performance of the safety system. (In this case, it must have a risk reduction factor [RRF] of 3,000.) The simple formula may be expressed as follows:

Required Risk Reduction Factor = Mean Time Between Hazard/Demand Rate

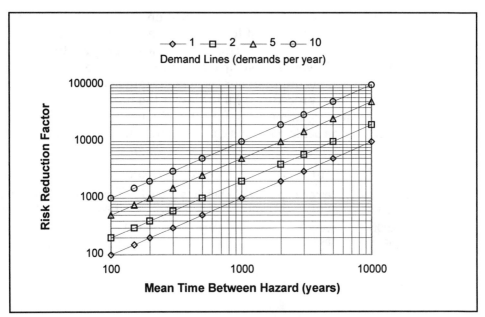

Figure 6-3   Risk Reduction Factor versus MTBH and Demand Rate

Although this method has the benefit of being purely quantitative (there is no "rare to frequent" open to subjective interpretation here), it requires stating a target goal, which some are unable, or unwilling, to do. This presents legal problems, especially in the United States. If one chooses a target of one thousand years between a hazard and an accident occurs, the lawyers will probably ask, "Why didn't you choose ten thousand years?" Recall from Chapter 2 that a target of one thousand years does not mean the hazard will not occur for one thousand years. It means that if there were one thousand installations, in one year one of them might have the hazard.

6.6.1   Layers of Protection Analysis (LOPA)

A variation of the quantitative method is described in the AIChE CCPS text as the layers of protection analysis (LOPA). The required risk reduction factor formula shown in the previous section implies that there is only one safety layer in the plant to prevent the hazard. This is rarely the case. Figure 4-1 in Chapter 4 shows how there are multiple layers of protection in any process plant. Each layer can have its own associated level of performance or risk reduction.

For example, the overall goal might be a probability of multiple deaths outside the plant of no more than one in ten thousand per year. (Some might say once in ten thousand years, but that is not exactly the same thing.) Because of the overall design of the basic process, there might be the possibility of an overpressure condition once every three days. The process control system might lower that to a hazard once per month. The alarms and operators might further lower the hazard rate to once per year. The SIS might lower the hazard rate to once per one hundred years. Mechanical systems might lower the hazard rate to once per one thousand years. Evacuation procedures might lower the hazard rate to once per ten thousand years. So, accounting for all of the protection layers, the overall risk level may be achieved. This can be shown mathematically as follows:

10,000	=	.01 x	10 x	10 x	100 x	10 x	10
overall safety goal		demand rate	process control	operators	SIS	mechanical systems	evacuation procedures
1 in 10,000 per year		every 3 days	monthly	yearly	1 in 100	1 in 1,000	1 in 10,000

Summary

Many industries need to evaluate and rank risk. Several techniques may be used, including both qualitative and quantitative. All methods involve evaluating the two components of risk (probability and severity). Of all the different methods, none is more "correct" than any other—all are merely approximations.

The task of performing a risk analysis and matching risk to safety system performance is not solely the responsibility of the control system engineer. It is the responsibility of a multidisciplinary team. Also, one should not try and determine the SIL for an entire process unit but rather determine the SIL for each safety function.

References

1. *Standard Practice for System Safety Requirements*, US MIL STD 882, ([location]: [abbreviation of organization], 19[year]).

2. ANSI/ISA, *Application of Safety Instrumented Systems for the Process Industries*, ANSI/ISA S84.01-1996 (Research Triangle Park, NC: ISA, 1996), ISBN 1-55617-590-6; American Institute of Chemical Engineers, *Guidelines for Safe Automation of Chemical Processes* (New York: AIChE, Center for Chemical Process Safety, 1993), ISBN 0-8169-0554-1; and International Electrotechnical Commission, *Functional Safety: Safety-Related Systems*, draft standard 61508 (Geneva, Switzerland: IEC, 1997).

7

CHOOSING A TECHNOLOGY

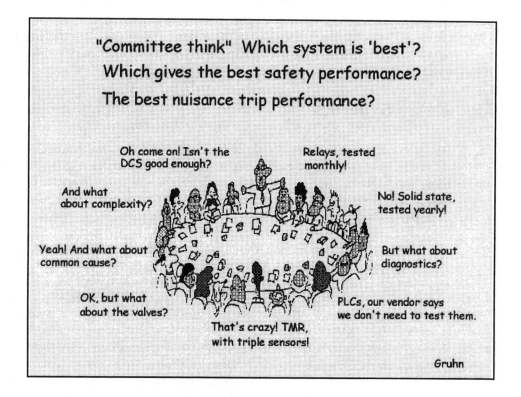

"If architects built buildings the way programmers write software, the first woodpecker that came along would destroy civilization."

— Unknown

A number of technologies are available for use in safety-instrumented systems: pneumatic, electromechanical relays, solid state, and PLCs (programmable logic controllers). There is no one overall "best" system; each has advantages and disadvantages. The decision over which system may be best suited for an application will depend upon many factors, such as budget, size, level of risk, flexibility, maintenance, interface and communication requirements, security, and the like.

7.1   Pneumatic Systems

Pneumatic systems are still in use and are perfectly appropriate under certain circumstances. A very common application for pneumatic systems has been the offshore industry, where many systems are required to operate without electrical power. Pneumatic systems are relatively simple (assuming they are "small") and relatively fail-safe. (Fail-safe in this sense means a leak would result in the system depressurizing, which would result in a shutdown.) These systems are most suitable for small applications where there is a desire for simplicity and intrinsic safety and where electrical power is not available.

7.2   Relay Systems

Relay systems are relatively simple (at least when they are small). They are relatively inexpensive to purchase, but because of nuisance trips they may have the highest cost of ownership. They are immune to most forms of electromagnetic interference (EMI) and radio frequency interference (RFI) and can be built for many different voltage ranges.

The larger relay systems get, the more unwieldy they become. Any time logic changes are required, wiring must be changed and drawings must be updated. If you have a relay panel that is a decade old or more, try this simple test. Grab the engineering logic drawings and walk out to the panel. Check to see if the two match. You may be in for a bit of a surprise.

The problems associated with changing wiring and documentation were some of the difficulties faced by the automotive industry. You can just imagine the thousands of relay panels they needed to produce all of the different automobile models and the constant changes that were required.

A not so obvious problem with relay systems are the wiring connections. One of the authors of this book has been to one facility where all the operators learned through "the school of hard knocks" not to even open the doors of their relay panels. That simple act has shorted out loose connections and caused plant shutdowns in the past.

Relay systems offer no form of communication with other systems (other than repeat contacts). Essentially, they're deaf, dumb, and blind. Also, most relay systems do not incorporate a way to test or perform bypasses. These features may be added, but the size and cost of the panels increase significantly.

Relay systems are based on discrete (on/off) logic signals. Traditionally, discrete input sensors (switches) were used. Analog signals could only be incorporated through the use of trip amplifiers, which provide a discrete output once an analog set point has been exceeded. Trip amplifiers, however, do not offer the same level of inherent fail-safe design as relays.

About the only time people now choose to use relays is for very small systems, typically those with less than about fifteen inputs and outputs (I/O). Relay systems are safe and can meet SIL 3 performance requirements, assuming the appropriate relays are used and the system is designed properly.

7.3 Solid-State Systems

Two different types of solid-state systems are currently available. One is a European system that is referred to as "inherently fail-safe." Similar to a relay, it has a known and predictable failure mode. (Not 100 percent of the system is truly fail-safe, however.) The other systems could be categorized as "conventional" solid state. They have more of a fifty-fifty failure mode split characteristic. About the only systems of this type that have survived to this day were originally developed specifically for safety applications and incorporate thorough diagnostics so as to find dangerous (inhibiting, fail-to-function) failures.

Solid-state systems are hardwired, much like relays. Opening some of these panels and looking at all of the wiring inside can be a bit intimidating for some. In addition to changing wiring, drawings must also be updated to reflect any modifications made to system logic.

The solid-state systems that were built for safety generally include features for testing and performing bypasses. Most of the systems also offer some form of serial communications to external computer-based systems. Solid-state systems perform simple relay logic; they are not capable of analog or PID control (although most are now able to accommodate analog input signals). This prohibits them from being used in certain applications that require math functions.

What may surprise many is the cost of these systems. Hardwired solid-state systems can be the most expensive solution; some are even more costly than triplicated PLCs. One might then ask, "Why use them at all?" The answer is simple. Some people want more functionality than relays but do not want a safety system that relies on software.

Whichever solid-state system one may consider, they all offer similar construction (see Figure 7-1). There are individual modules for inputs, outputs, and logic functions. (They do not use, for example, sixteen-channel input and output cards.) The modules must be wired into the logic configuration that is required for the system. This may involve literally miles of wire in some systems. When you stop to consider all of the detail engineering and manufacturing required, it is no wonder these systems are so expensive.

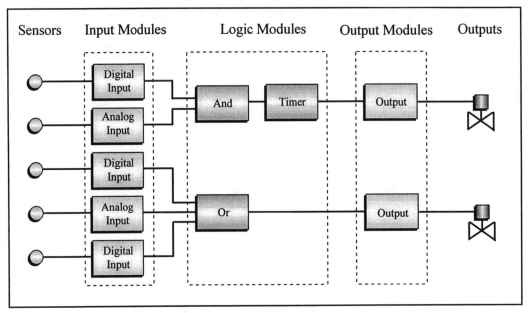

Figure 7-1   Typical Solid-state System

These systems offer several significant benefits over PLC-based systems. The most obvious is that they do not use software. Wiring is relatively easy to test and check. (Software, however, is difficult to verify under all circumstances. Do you ever really expect there to be such a thing as "bug-proof" software?) Since these systems do not "scan" the I/O the way a PLC does, these systems can respond faster than software-based systems (not getting so picky as to consider things like input filter delays, etc.).

These systems are usually supplied in a simplex, nonredundant configuration (although redundancy can be implemented). They do not have the potential for the systemwide problems typically associated with some software-based systems. In other words, single module failures will have a localized effect only for that single safety function. In contrast, in some PLCs, if you lose the CPU you lose everything.

7.4 Microprocessor/PLC (Software-based) Systems

Computer-based systems appear to be the system of choice for many applications today. There is no technological imperative, however, that says we *must* use computers.

PLCs (programmable logic controllers) were originally designed to replace relays, yet their application as shutdown systems requires close scrutiny. They offer low cost, the ability to make changes easily and flexibly, serial reporting capabilities, graphical operator interfaces, self-documenting features, and the like. PLCs, however, were not designed for use in critical safety applications, and most units do not have extensive diagnostic capabilities, fail-safe characteristics, or effective levels of redundancy. Quantitative analysis of such systems shows that most units are generally not suited or recommended for use in SIL 2 and 3 applications.

7.4.1   Flexibility: Cure or Curse?

Flexibility offers benefits for some but introduces other problems. The ease with which changes can be made will encourage some to actually make changes, which can often lead to complexity and introduce new errors. Flexibility encourages the redefinition of tasks late in the development phase (or after installation) so as to overcome deficiencies found elsewhere in the system. As John Shore put it, "Software is the place of afterthoughts." This flexibility also encourages premature construction. Few engineers would start to build a chemical plant before the designers had finished the detailed plans. This flexibility also allows the use of unique and sometimes unproven techniques. Similarly, few engineers would design a complex system, such as a jet aircraft, after having only built a model.

Hardwired systems impose physical limitations on the design of a system. This helps to control complexity. (You can only connect so many wires, timers, and relays in a given circuit.) In contrast, software has no corresponding physical limits. It therefore becomes possible to build enormously complex software-based systems.

These systems are "modular" in design; therefore changes can be made quickly and simply (compared to hardwired systems). PLCs tend to have a relatively low cost and the smallest overall footprint. They offer serial communication links to other systems and are self-documenting.

The level of testing and diagnostics required by these units varies considerably. Since most PLCs were designed for active, dynamic control, they do not have a need for extensive diagnostics (which only increase the cost). This lack of diagnostics tends to be the weak link in these sys far as safety is concerned, as we shall see.

7.4.2   Software Issues

Just as some list software as the primary strength of PLCs, others view this "dependence" upon software as the biggest weakness. This vulnerability involves two areas, reliability and security. How many programmers when they are through with a project can walk into their boss's office, put their hands over their hearts, and claim, "Boss, it's going to work every time, I guarantee it"? Also, most programmable systems have very little in terms of access security.

According to one estimate, errors in the systems software (the software supplied with the computer) can vary between one error in thirty lines at worst and one error in a thousand lines at best. Michel Gondran quotes the following figures for the probability that there will be a significant error in the applications software of a typical microprocessor-based control system:[2]

> Normal systems: 10^{-2} to 10^{-3}.

> To achieve 10^{-4} considerable effort is needed.

> To achieve 10^{-6} the additional effort has to be as great as the initial development effort.

Another issue, generally only associated with redundant systems, is common cause. Common cause is a single stressor or fault that can make redundant items fail. Software is one such potential problem. If there is a bug in the software, the entire redundant system will not operate properly.

7.4.3   Hot Backup PLCs

Although many people do use simplex PLCs for safety-related applications, most critical systems employ some form of redundancy. One of the most popular schemes is a hot backup system (see Figure 7-2). This system employs redundant CPUs (although only one is on line at a time) and simplex I/O modules. I/O modules generally communicate with the CPUs through some form of communication module. These communication modules generally have no provision for redundancy. A number of items are worth pointing out about such hot backup systems.

For the system to switch to the backup CPU, it must detect a failure of the online unit. Unfortunately, this does not happen 100 percent of the time because CPU diagnostics cannot detect 100 percent of all possible failures. A general figure is about 80 percent diagnostic coverage, possibly rising as high as 95 percent if a number of additional features are implemented (e.g., watchdog timers).

Most PLC vendors assume the switching is 100 percent effective. This is obviously an idealistic assumption, as a variety of switch failures have been reported. Some switches must be programmed in order to work.

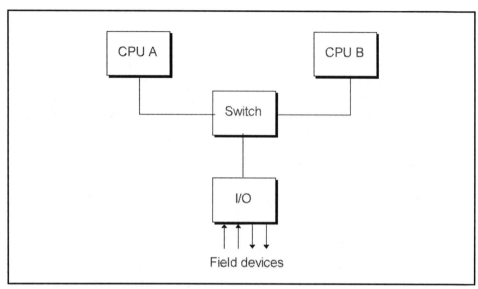

Figure 7-2   Hot Backup PLC System

The real weak link, however, in most general-purpose PLCs is the potential lack of diagnostics for the I/O modules. Some units literally have no diagnostic capabilities at all. (Users are urged to ask their vendors for this sort of information. According to ANSI/ISA S84.01-1996, the vendors are supposed to provide it.) There are some units with relatively good diagnostics. When asked why more people aren't using them, the typical response is "They're too expensive." No kidding! You get what you pay for.

A number of rather frightening stories have been told about such systems when used in safety-related applications. Unfortunately, they are not widely publicized because of their sensitive nature. For example, one of this book's authors knows an engineer who told how his company was considering using a PLC for safety. They put a demo system through a number of tests. One test was to take a running system with all of the I/O energized and remove the CPU chip *while the system was running*. They did just that, and the system *did not respond* at all to the change—all of the I/O stayed energized, and there were no alarms or indications of any kind! He phrased it rather well: "We gave the system a lobotomy, and it didn't even recognize anything happened!" Then he added: "We obviously didn't use that system."

One facility had seven systems like this for safety-related applications. The users heard some troubling stories, so they went out and tested the units. They "tickled" an input and checked for the appropriate output response. Of the seven systems they tested, four did not respond properly! What is alarming is that no one knew anything was wrong until they actually went out and tested the systems. All the lights were green, so they naturally assumed everything was OK.

One engineer at a corporate office reported that he sent out a memo to all of his companies' plant sites requesting that the PLCs they used for safety be tested. The reports he got back said that between 30 percent and 60 percent of the systems did not respond properly. Again, this does not make headline news, and this is a very sensitive topic. No one presents a paper at an ISA conference saying, "50 percent of our safety systems don't work. Boy, did *we* screw up!"

One engineer reported that his engineering and construction (E&C) company designed a hot backup PLC system per the customer's specifications. They installed it, tested it, and everything worked fine. They tested it a year later, and everything still worked fine. Then one day the PLC shut the unit down. When they checked the system, they found that the online unit failed, and the system switched to the backup, but the backup CPU cable had never been installed! The system ran for one and a half years and could not detect the condition, and no one in the plant or with the contractor realized the situation.

7.4.4   Dual Redundant General-purpose PLCs

An approach tried by some is to take off-the-shelf hardware and modify the system (i.e., customize it) to try and provide a higher level of diagnostic coverage. One such scheme involves the addition of I/O modules, loop back testing, and additional application programming to make all this happen (as shown in Figure 7-3).

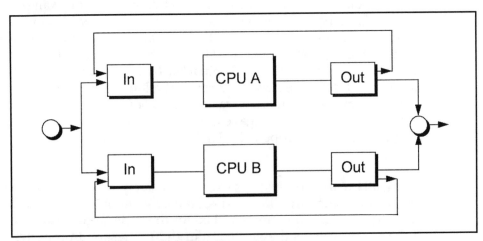

Figure 7-3   **Active Parallel Redundant System**

Unfortunately, systems of this type tend to be large and "special" and utilize programming that some consider difficult to maintain. When one considers the extra hardware, programming, and engineering effort, this can become rather expensive. A word used by some (which others might be offended by) would be *kludge*. One of the authors has known some who tried this approach and then said, "Never again!" This "special" dual approach might have made sense in 1980, but it is not very economical when compared with the standard systems that are now available.

There is an inherent problem with any dual system of this type. When the two channels don't agree, which one is right? Is one channel experiencing a safe (initiating) failure (and you do *not* want to trip, so you choose two-out-of-two [2oo2] voting)? Or is one channel not responding due to a dangerous (inhibiting) failure, and the alarm is genuine (and you *do* want to trip, so you choose one-out-of-two [1oo2] voting)? Again, which one is right? With most standard off-the-shelf systems this is very difficult to determine.

7.4.5   TMR (Triple Modular Redundant) PLCs

Triple modular redundant (TMR) systems are essentially specialty triplicated PLCs (see Figure 7-4). The original research for these systems was funded by NASA in the 1970s. The systems that were developed commercially offer the same benefits described earlier for PLCs yet were specifically designed for safety applications and incorporate extensive redundancy and diagnostics. With triplicated circuits, such a system can survive single (and sometimes multiple) safe or dangerous component failures (hence the term *fault-tolerant*). These systems are suited for use in all SIL applications.

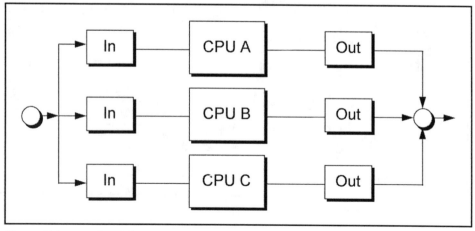

Figure 7-4   TMR System

The diagram in Figure 7-4 is generic. Some of the systems utilize "quad" redundant outputs. Some employ redundant circuits on one board (module); others utilize multiple modules.

In general, these systems do not require any additional overhead programming in order to accomplish diagnostics. (There are exceptions, however.) The user only writes and loads one program, not three separate ones. (Again, there are exceptions.) Essentially, the triplication is designed to be transparent to the user.

7.4.6   Dual 1oo2D PLCs

Since the late 1980s a number of dual systems, termed *1oo2D*,[3] were designed specifically for the process industry. These systems are designed to be fault-tolerant to the same degree as triplicated systems–they require two simultaneous failures to fail either safely or dangerously. They are based on more recent concepts and technology and are certified by independent agencies such as TÜV (Technisher Überwachungsverein) and FM (Factory Mutual) to the same safety levels as the triplicated systems. In addition to dual configurations, these vendors can also supply simplex (nonredundant) systems that are still certified for SIL 2 applications. In other words, these systems are also suited for use in all SIL applications. An example of such a system is shown in Figure 7-5.

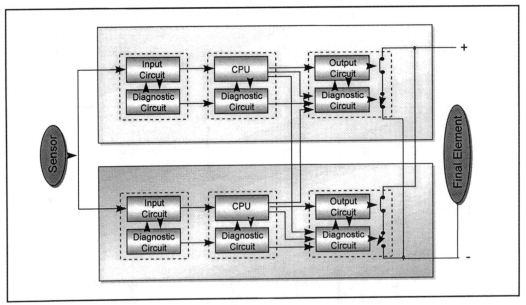

Figure 7-5   1oo2D System

7.5 Issues Related to System Size

The larger a system gets, the more complicated it becomes. Size also has an impact on modeling the performance of a system. (Refer to Chapter 8 for more information on modeling.)

In the past, most relay systems were small and distributed around the plant. Each process unit or piece of equipment typically had its own dedicated shutdown system. As the control systems in most plants became centralized, the shutdown systems also became more centralized. Rather than keeping multiple, small, hardwired systems, many systems migrated into single, centralized, larger units. A small relay panel is relatively easy to manage and maintain. The same cannot be said of a five-hundred-I/O relay system. Hardwired systems were thus frequently abandoned in favor of smaller, easier-to-manage, software-based systems. This centralization introduces a new set of problems, as single failures now may have a much wider impact. Single failures in small segregated systems will have a limited, localized impact only. Single failures in large centralized systems, however, may have a major impact. For example, if a simplex PLC is controlling two hundred I/O and the CPU fails, one loses control of all two hundred I/O. For this reason, many software-based systems are designed with some form of redundancy. The drawback to redundancy is that it usually results in additional system complexity.

7.6 Issues Related to System Complexity

One way to deal with complexity is to break the design (e.g., software) into smaller pieces or modules. Although this can reduce the complexity of the individual components, it increases the number of interfaces between all of the components. This can have the detrimental effect of actually increasing the overall complexity. It becomes difficult to comprehend the many conditions that can arise through the interactions of all of the components.

In terms of safety instrumented systems (SIS), simpler is better. Instead of spending money for extra complexity, it may be better in the long run to spend money for simplicity. Simpler systems are easier to analyze, design, build, test, and maintain.

Complexity can be included when modeling the performance of systems. A functional or systematic failure rate can be incorporated in the calculations. This may be a design, programming, fabrication, installation, or maintenance error. Although the exact number chosen may be rather subjective and difficult to substantiate, it is relatively easy to include in a model. (Refer to Chapter 8 for an example of a model.)

7.7   Communications with Other Systems

Most plants now have some form of centralized control system with graphical operator displays. Rather than having a separate set of displays for the safety instrumented system, there are benefits to having the SIS information available for display at the main control system consoles (e.g., lower cost, less space, learning only one system, etc.). This usually requires some form of serial communication link between the systems.

Most computerized systems offer some form of serial communications. It is relatively easy to have the process control system read or poll the SIS and display information such as the state of all the I/O, bypasses, alarms, and the like. It is just as easy to have the process control system write to the SIS, and herein lies the potential danger. The control system must not be allowed to corrupt the memory of the SIS. This requires some form of control. In other words, the control system can read information about the SIS, but one should be very careful about writing information to the SIS. There will no doubt be certain instances where the control system needs to write to the SIS. This should be carefully considered and carefully controlled. Consider read-back testing to make sure only the intended variable has been altered.

Summary

A number of technologies are available for use in safety instrumented systems: pneumatic, electromechanical relays, solid state, and PLCs (programmable logic controllers). There is no one overall "best" system; each has advantages and disadvantages. The decision on which system may be best suited for an application will depend upon many factors, such as budget, size, level of risk, flexibility, maintenance, interface and communication requirements, security, and so on.

Pneumatic systems are most suitable for small applications where there are concerns over simplicity, intrinsic safety, and a lack of available electrical power. Relay systems are fairly simple, are relatively inexpensive to purchase, are immune to most forms of EMI/RFI, and can be built for many different voltage ranges. They generally do not incorporate any form of interface or communications. Changes to logic require manually changing documentation. In general, relay systems are used for relatively small applications.

Solid-state systems (hardwired systems that do not incorporate software) are also available. Several of these systems were built specifically for safety applications and include features for testing, bypasses, and communications. Logic changes still require manually changing documentation.

These systems have fallen out of favor with many because of their limited flexibility and high cost as well as the acceptance of software-based systems.

Software-based systems, generally industrial PLCs (programmable logic controllers), offer software flexibility, self-documentation, communications, and higher-level interfaces. Unfortunately, many general-purpose systems were not designed specifically for safety and do not offer features required for many applications. However, certain specialized dual and triplicated systems were developed for the more critical applications and have become firmly established in the process industry.

References

1. J. R. Taylor, "Safety Assessment of Control Systems: The Impact of Computer Control." Paper presented at the Israel Institute for Petroleum and Energy Conference on Process Safety Management, Tel Aviv, Israel, October 1994.

2. Michel Gondran, remarks quoted at launch meeting of the European Safety and Reliability Association, Brussels, Belgium, October 1986.

3. William M. Goble, *Control Systems Safety Evaluation & Reliability* (Research Triangle Park, NC: ISA, 1998), ISBN 1-55617-636-8.

INITIAL SYSTEM EVALUATION

"A failure once in a million years?! Common sense leads me to think you've neglected some factors.

Gruhn

"There's always an easy solution to every human problem, neat, plausible... and wrong."

— *H. L. Mencken*

8.1 Things Are Not As Obvious As They May Seem

If it were intuitively obvious which system was most suitable for a particular application, then there would be no need for this book. The problem is that things are *not* as intuitively obvious as they may seem. Dual is *not* always better than simplex, and triple is *not* always better than dual. Consider the choices presented in Table 8-1.

Sensors	Logic	Diagnostic Coverage	Common Cause	Outputs	Test Interval
Single	Single	99.9%	N/A	Single	Monthly
Dual	Single	99%	N/A	Dual	Quarterly
Triple	Single	90%	N/A	Dual	Yearly
Single	Dual	99%	0.1%	Single	Monthly
Dual	Dual	90%	1%	Dual	Quarterly
Triple	Dual	80%	10%	Dual	Yearly
Single	Triple	99%	0.1%	Single	Monthly
Dual	Triple	90%	1%	Single	Quarterly
Triple	Triple	80%	10%	Single	Yearly

Table 8-1   **Which System Is "Best"? Which Gives the Fewest Nuisance Trips? Which Gives the Best Safety Performance?**

Consider just the nine different cases shown in Table 8-1, all software-based systems. Let's not even concern ourselves at the moment with relay and solid-state systems.

First, remember that safety instrumented systems (SIS) can fail in two ways. They may suffer nuisance trips and shut the plant down when nothing is actually wrong, or they may fail to function when they are actually required. Could one system work best in both modes? If a system is good in one mode, would it necessarily be bad in the other?

A chain is only as strong as the weakest link. Should the sensors be redundant? Dual redundant devices can be configured in at least two ways: one-out-of-two (1oo2) or two-out-of-two (2oo2). In other words, will the system trip when only one sensor goes into alarm or only when both go into alarm?

Should the logic solver be redundant? There are at least four different ways of configuring dual logic solvers. What about triplicated?

Diagnostic coverage means the percentage of failures that can be detected automatically by the system. Do not assume that modern electronic devices have 100 percent diagnostic coverage. The example shown in Table 8-1 is not meant to imply that there is a simplex programmable logic solver with 99.9 percent diagnostic coverage; it is just an example. With that in mind, would a simplex system with 99.9 percent diagnostic coverage be "better" than a triplicated system with 80 percent coverage?

Common cause means a single stressor or fault that impacts an entire redundant system. Typical examples are external environmental stresses to the system, such as heat, vibration, overvoltage, and the like. One method of quantifying common cause is referred to as the beta factor. This represents the percentage of failures that can impact multiple channels at once. For example, if a redundant system has 1 percent common cause-related problems, it means that of all the failures identified, 1 percent of them might hit multiple channels at the same time and make the entire system fail. Is 1 percent enough to worry about? What about 10 percent?

Should the final elements (valves) be redundant? In dual or in series? This option gets rather expensive, but is it important? Triplicated outputs are not shown because it is not possible to connect three valves and get two out of three voting.

How often should the entire system be manually tested (for faults the system cannot diagnose on its own)? Should everything be tested at the same interval, or might different portions be tested at different intervals? Do redundant systems need to be tested more often, or less often, than simplex (nonredundant) systems?

Refer back to Table 8-1 and decide. Which one is best? Will your gut feeling, intuition, or experience give you the same answer as someone else? Probably not.

Various standards groups have been meeting for years trying to resolve these issues. If all of this were "easy," the groups would have been done *long* ago! Consider the ISA SP84 committee alone. There were over three hundred members on the committee mailing list, representing about every interest group imaginable. How can such a group reach agreement on a topic as complicated and controversial as this? How can one peer past all the hype?

Figure 8-1   Consensus by Committee

Intuition may be fine for some things, but not for others (see Figure 8-2). Jet aircraft are not built by "gut feeling," bridges are not built by "experience," and nuclear power plants are not build by "intuition." If you were to ask the chief engineer of the Boeing 777 why they used a particular size engine, how comfortable would you feel if their response was, "Well, we weren't sure ... but that's what our vendor recommended"?

Figure 8-2   **Intuition may be fine for some things, but not for others**

8.2   Why Systems Should Be Analyzed before They're Built

Since things are not as intuitively obvious as one may wish, it is important to be able to analyze systems in a quantitative manner. Although quantitative analyses may be imprecise (as will be stressed shortly), they nevertheless are a valuable exercise for the following reasons:

- They provide an early indication of a system's potential for meeting the design requirements.

- They enable life cycle cost comparisons.

- They enable one to determine the weak link in the system (and fix it, if necessary).

- They allow an "apples-to-apples" comparison between different offerings.

8.2.1   Caveats

"There are lies, there are damn lies, and then there's statistics."

— M. Twain

Simple system models may be calculated and solved by hand. As more factors are accounted for, however, manual methods become rather unwieldy. It is possible to develop spreadsheets or other computer programs to automate the process. A major drawback of some models is often not what they include, but what they do *not* include. One can model a triplicated system according to one vendor's optimistic assumptions and then model it with a more realistic set of assumptions and change the answer by four orders of magnitude! It is not the *accuracy* of the model that matters, it is the *assumptions* that go into it. After all, computers are known for their speed, not their intelligence.

"Computer models can predict performance with great speed and precision, yet they can also be completely wrong!"

— Unknown

"However neutral computers may be, they can never produce an objective answer from highly subjective data."

*— Imperato and Mitchell, **Acceptable Risks**, 1985.*

One needs to apply a bit of common sense to modeling. For example, if two events each have a probability of 10^{-6}, a simplistic approach would be to say that the possibility of the two events happening simultaneously would be 10^{-12}. Low numbers such as this simply mean the system will most likely fail in a way not considered. Absolutely absurd risk estimates based on failure rate data are not uncommon.[1] The current record is an estimate of a nuclear weapon system where an event probability was estimated at 2.8×10^{-397}. Considering that the probability of being struck by a falling plane is roughly 10^{-8}, the absurdity of such computations becomes obvious. Trevor Kletz phrases it rather bluntly: "One wonders how scientifically trained people can accept such garbage."[2]

"Reliability models are like captured foreign spies; if you torture them long enough, they'll tell you anything."

— P. Gruhn

In a related thought, Kletz has pointed out that "time is usually better spent looking for all the sources of hazard, than in quantifying with ever greater precision those we have already found." For example, in the space program, where quantitative fault tree analysis (FTA) and failure mode and effects analysis (FMEA) were used extensively, almost 35 percent of actual in-flight malfunctions had not been identified.[1]

8.3   Where to Get Failure Rate Information

To predict the performance of a system, one needs performance data on all the components. Where does one get this sort of information? The following sections discuss some sources.

8.3.1   Maintenance Records

Hopefully, each plant has maintenance records showing the failure rates of all the devices in the plant. The federal government's Process Safety Management (PSM) legislation (29 CFR 1910.119) implies that users should be keeping this sort of data. Many plants do in fact have the information, although it may not initially be collated in a usable fashion. Experienced technicians generally have a good idea how often they have to repair different pieces of equipment. But if one does not have such records, then what?

8.3.2   Vendor Records

One could ask vendors for their maintenance records, but feedback received from those who have tried is not very encouraging. Even if the vendor does provide data, it is important to ask how *they* got it. Is it based on field returns or indoor laboratory experiments? How many customers of a ten-year-old system actually send failed modules back to the factory? What about a brand-new system that has never even been built? One obviously cannot get field return information on that.

8.3.3   Third-Party Databases

Commercial databases for the offshore, chemical, and nuclear industries are available. This information has been compiled and made available for everyone's use in the form of several useful books.[3] This may actually represent the best sort of data (if a local plant does not have its own), because it is based on actual field experience from different users.

But what about equipment that is brand new and does not yet have a history of operating in the field?

8.3.4   Military-Style Calculations

Decades ago the military was faced with this problem of having brand-new equipment that did not yet have a history of field operation. When a nuclear submarine receives the launch code, one wants to know that the likelihood of the communications system working properly uses some measure other than just "high." The military developed a technique for predicting the failure rates of electronic systems (MIL-HDBK 217F).[4] This handbook has gone through many revisions and has long been the subject of much controversy. All who use it are aware that it tends to give pessimistic answers (sometimes by orders of magnitude). That does not mean, however, that it should not be used, just that it should be used with caution. It still provides an excellent yardstick for comparing systems, even if its absolute answers are questionable. The publishers of the document are aware of this and state the following: "a reliability prediction should never be assumed to represent the expected field reliability as measured by the user" (MIL HDBK 217F, paragraph 3.3).

8.4   Failure Modes

Many people have replaced relay systems with PLCs (programmable logic controllers). Their typical response as to why they did so is usually, "That's what a PLC was designed for—to replace relays." In terms of safety, however, that should *not* be the main issue. Obviously, a PLC can do everything a relay system could and a whole lot more (e.g., timers, math functions, etc.). With a safety system, the concern should not so much be how does the system *operate* but rather how does the system *fail*? This concept is so simple that it is typically overlooked, yet it is the underlying reason why safety-related systems differ from control systems and why safety instrumented systems have unique design considerations.

Safety instrumented systems can fail in *two* ways. First, systems may initiate nuisance trips. That is, they may shut the plant down when nothing is actually wrong. An example would be a closed and energized relay that just pops open. People have given these type of failures a number of different names: overt, revealed, fail-safe, and so on. Since they result in plant shutdowns, they tend to be costly. People want to avoid them for economic reasons. When systems suffer too many failures like this, people tend to lose confidence in them, and they are frequently bypassed as a result. One term for these sort of failures is "initiating," since they *initiate* action, whatever that action may be (open–closed, on–off, energized or deenergized). The industry seems to have settled on the term *safe failure* for this failure mode.

One must not forget, however, that systems may also suffer failures that will make them fail to respond to a true demand. Some have called these covert, hidden, fail-danger, and so on. Such names, however, do not imply what is really important– the system *will not function when required*. If a system fails in this manner (e.g., the relay contacts are welded shut) it would represent a potential danger, for if there were a demand the system would not respond. One term for these sort of failures is *inhibiting*, since the system is *inhibited* from responding. The industry seems to have settled on the term *dangerous failure* for this failure mode. The only way to find these failures (before it is too late) is to *test* for them. Many people do not understand the need for this. Remember, however, that safety instrumented systems are dormant, or passive, and failures are *not* inherently revealed. Unfortunately, many systems do not have effective diagnostics, a concept rarely discussed by most vendors.

Examples of hidden failures can be somewhat embarrassing and are therefore often "buried." One example was a satellite manufacturer that used a crane to lift a $200,000 satellite into its shipping container. When they raised the satellite, the crane was unable to stop the upward motion. When the satellite reached the top of the crane, the cable snapped and the satellite crashed to the floor. The relay responsible for controlling the upward motion was replaced with one rated for only 10 percent of the actual load, so the contacts welded. The crane was leased and was not tested before being used.[11]

8.5   Metrics

To measure and compare the performance between different systems, one needs a common frame of reference, a common *understandable* set of terms. Many performance terms have been used for years, such as availability and reliability, yet they have caused problems. If you ask four people what these terms mean, you will no doubt get four slightly different answers. The main reason for this stems from the two different failure modes. If there are two failure modes, there should be *two* different performance terms, one for *each* failure mode. How can one term, such as *availability*, be used to measure the performance of two different failure modes?

Another reason the term *availability* causes confusion is the typical range of numbers encountered. Anything over 99 percent sounds impressive. Whenever PLC vendors give performance figures, it always ends up being a virtually endless string of nines. Stop and consider whether there is an appreciable difference between 99 percent and 99.99 percent availability. It's less than 1 percent—how much of a difference can there be?

In terms of safety, some prefer to use a measure related to availability called the probability of failure on demand (PFD). Unfortunately, the numbers are difficult to relate to because they are so small that one needs to use scientific notation. If the purpose of this exercise is to give management meaningful information so it can make intelligent decisions, will telling your supervisor, "Hey boss, our safety system has a PFD of 2.3×10^{-4}!" be helpful? No doubt the manager's first response will be "Uh, is that good?"

For example, consider the pie chart shown in Figure 8-3. The upper right third is where the safety system is performing properly and the plant is up and producing product. In this case, the safety instrumented system (SIS) and the plant are both available.

In the lower section of Figure 8-3, the SIS has suffered a nuisance trip and shut the plant down. In this case both are unavailable. (Granted, you do have to get the SIS up and running before you can get the plant started, so the two are not exactly the same, but that's getting picky.) However, the case most people are not even aware of is the upper left portion of the figure, where the SIS is nonfunctional (e.g., a PLC stuck in an endless loop, outputs energized and stuck on, etc.), so the SIS is *not* available, but the plant *is* still up and running. This sort of system failure does not shut the plant down, so the plant is available and still making product (and oblivious to the fact that the SIS will not function when needed). In this case, the "availability" of the two systems is clearly *not* the same.

This also shows that you should not combine the two different measures of performance with one term (such as an "overall" availability) because the impact of each failure is completely different. It hardly makes sense to take an average of 99.999 percent in the safe mode and 94.6 percent in the dangerous mode. What would the final answer tell you? Absolutely nothing.

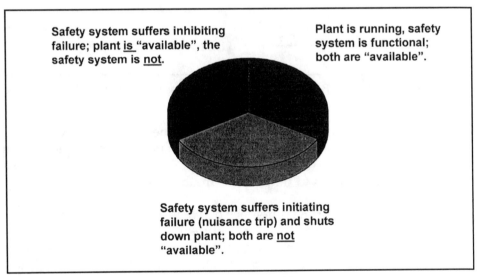

Safety system suffers inhibiting failure; plant is "available", the safety system is not.

Plant is running, safety system is functional; both are "available".

Safety system suffers initiating failure (nuisance trip) and shuts down plant; both are not "available".

Figure 8-3   **Confusion with the Term** *Availability*

8.5.1   Failure Rate, MTBF, and Life

The bathtub curve (see Figure 8-4) illustrates that failure data is not fixed over the entire life of a device. The curve shown in Figure 8-4 is generally accepted for electronic devices. Software and mechanical devices tend to have slightly different curves.

The left portion of the curve shows the impact of "infant mortality" failures. In other words, defective components die young. The right portion of the curve shows "wear-out" failures. Devices can only last so long, hence the curve at some point returns to zero (as shown by the dotted portion of the line in Figure 8-4). For example, not many people live past one hundred years of age.

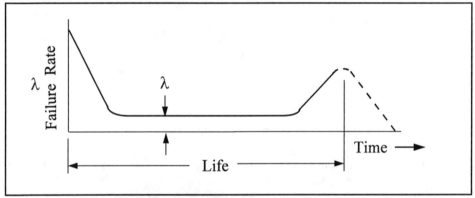

Figure 8-4   **Bathtub Curve Illustrating Life, Failure Rate, and MTBF (1/λ)**

A constant failure rate is generally assumed for most devices. This is represented by the middle portion of the curve. This simple assumption is still a topic of heated debate in the reliability community. But the assumption at least tends to simplify the math involved.

Some argue over the difference between MTTF (mean time to failure) and MTBF (mean time between failures). Traditionally, MTTF is used for replaceable components (e.g., resistors), and MTBF is used for repairable systems (e.g., aircraft). For the sake of consistency, MTBF is used in this chapter.

Many intermix MTBF and life. The classic example of how MTBF and life are *not* the same is a simple match. When using dry matches and the proper technique, there will be few failures. Therefore, the failure rate (failures per unit time) will be low. If the failure rate is low, the reciprocal, MTBF, will be high, or large. But a match only burns for a few seconds. The distinction between life and MTBF must be stressed. An MTBF of three thousand years may be a perfectly valid number for some components, even though they obviously will not last for three thousand years. It merely means that out of a total of three thousand devices, in one year one typically fails.

8.6 Degree of Modeling Accuracy

All reliability analyses are based on failure rate data. It must be recognized that such data is highly variable. Allegedly identical components operating under supposedly identical environmental and operating conditions are not realistic assumptions. For a given level of detail, the apparent precision offered by certain modeling methods is not compatible with the accuracy of the failure rate data. As a result, it may be concluded that simplified assessments and the use of relatively simple models will suffice. More accurate predictions can be both misleading and a waste of time, money, and effort. In any engineering discipline the ability to recognize the degree of accuracy required is of the essence. Since reliability parameters are of wide tolerance, judgments must be made on one or at best two-figure accuracy. Benefits are obtained from the judgment and subsequent follow-up action, not from refining the calculations.[5] Simplifications and approximations are useful, however, when they reduce complexity and allow a model to become understandable.[6]

8.7 Modeling Methods

"Managing lots of data begins with some form of simplification."
                                                                                  — *Megill*

"The less we understand a phenomenon, the more variables we require to explain it."
                                                                                  — *L. Branscomb*

A number of methods are available for estimating the performance of systems. Some of the more common are reliability block diagrams, fault trees, and Markov models.

Reliability block diagrams, along with their associated simple series/parallel formulas, can be found in most any reliability textbook. Block diagrams are just that, *diagrams* that help clarify system configuration and operation. The formulas typically used are simple but in general do not handle time-dependent variables, such as repair times, test intervals, diagnostics, and the more complex redundant systems. (A two-out-of-three [2oo3] system can be represented by a series of AND and OR gates or a set of series/parallel formulas, but that is getting a bit convoluted.)

Fault trees are excellent for modeling entire systems, including the field devices and the process (as we shall see). Fault tree diagrams are based on combinations of AND and OR gates and may quantitatively be solved by multiplying or adding the probabilities of the inputs to the gates. They are of limited value in modeling logic systems, however, because, much like block diagrams, they do not account for time-dependent factors. In addition, fault trees only account for "known" events. In other words, if you are not even aware of a particular failure, you subsequently cannot include it in the fault tree.

Most reliability practitioners have settled upon Markov models. Markov modeling is typically associated with complex transition diagrams and matrix math (although algebraic simplifications have been available for decades). The methods can be quite flexible; however, few understand how to solve them.

8.7.1   Basic Formulas

The theory behind the following formulas is developed in *Reliability, Maintainability, and Risk* by David J. Smith.[5] The formulas are based on algebraic simplifications of Markov models. For safe (initiating) failures, the formulas for calculating the mean time between nuisance trips are as follows:

Formula Set 1: (nuisance trip calculation)

    1oo1 $1 / \lambda_s$

    1oo2 $1 / (2 * \lambda_s)$

    2oo2 $1 / (2 * (\lambda_s)^2 * MTTR)$

    2oo3 $1 / (6 * (\lambda_s)^2 * MTTR)$

where:

    MTTR =   mean time to repair,

    λ =   failure rate (1 / MTBF)

    s   =   safe (initiating) failure

("1oo1" stands for "1 out of 1"; "2oo3" stands for "2 out of 3"; etc.)

These formulas are valid when the repair rate is much greater than the failure rate (1/MTTR >> λ). It is assumed that safe failures are revealed in all systems, even 2oo2 and 2oo3 configurations (through some form of discrepancy alarm).

For inhibiting (dangerous) failures, the unavailability (PFD_{avg}—average probability of failure on demand, or FDT—fractional dead time) is as follows:

Formula Set 2: (PFD_{avg} calculation for detected failures)

1oo1 $\lambda_d * (MTTR + (TI_a / 2))$

1oo2 $2 * (\lambda_d)^2 * (MTTR + (TI_a / 2))^2$

2oo2 $2 * (\lambda_d) * (MTTR + (TI_a / 2))$

2oo3 $6 * (\lambda_d)^2 * (MTTR + (TI_a / 2))^2$

where:

TI_a = Automatic diagnostic interval (which is usually insignificant)

MTTR = mean time to repair

λ = failure rate (1 / MTBF)

d   = dangerous (inhibiting) failure

These formulas assume that repairs are made as soon as failures occur. This is generally valid with "detected" failures but is not valid in systems with periodic manual testing. In that case, the repair time is negligible compared to the inspection interval (hours versus months), so the PFD formulas become the following:

Formula Set 3: (PFD_{avg} calculation for undetected failures)

1oo1 $\lambda_d * (TI / 2)$

1oo2 $((\lambda_d)^2 * (TI)^2) / 3$

2oo2 $\lambda_d * TI$

2oo3 $(\lambda_d)^2 * (TI)^2$

where:

TI   = Manual test interval

One needs to first split the component failure rates into the safe and dangerous failure modes. The first set of formulas is used to derive the system mean time between nuisance trips, or the nuisance trip rate. The second and third set of formulas are used to derive the PFD_{avg} by splitting the calculation into *nine* parts. For example, assuming a 2oo3 configuration, these nine parts are as follows:

(a) 2oo3 portion detected with automatic diagnostics

(b) 2oo3 portion detected with manual testing

(c) 2oo3 portion never detected

(d) common cause (beta) portion detected with automatic diagnostics

(e) common cause (beta) portion detected with manual testing

(f) common cause (beta) portion never detected

(g) common cause (systematic) portion detected with automatic diagnostics

(h) common cause (systematic) portion detected with manual testing

(i) common cause (systematic) portion never detected

One may assume the same automatic and manual coverage factors for common cause (beta and systematic—[d] through [i]) failures as for the formulas for parts (a) through (c). In other words, if the hardware has 80 percent automatic diagnostic coverage, then 80 percent of common cause failures will be detected by automatic diagnostics. If the hardware has 90 percent manual test coverage, then 90 percent of the remaining common cause failures will be detected through manual testing.

The failures detected automatically use the formulas in Formula Set 2, where the automatic diagnostic interval is negligible compared to the repair time (seconds or minutes versus hours). The portion of failures detected manually uses the formulas in Formula Set 3, with TI as the periodic manual test interval. The portion never tested also uses the formulas in Formula Set 3, but now TI is the proposed life of the system. In other words, faults might exist undetected for the life of the system.

8.7.2 Example

Using Ca for the automatic coverage factor and Cm for the manual coverage factor, a 2oo3 example (not getting into module quantities just yet) with the nine portions listed in the last section (in the same order) would be as follows:

$$6 * (\lambda_d * Ca)^2 * (MTTR)^2 +$$

$$(\lambda_d * (1 - Ca) * Cm)^2 * (TI)^2 +$$

$$(\lambda_d * (1 - Ca) * (1 - Cm))^2 * (Life)^2 +$$

$$(\lambda_d * Ca * (beta) * MTTR) +$$

$$(\lambda_d * (1 - Ca) * Cm * (beta) * (MTTR + TI/2) +$$

$$(\lambda_d * (1 - Ca) * (1 - Cm) * (beta) * (Life) +$$

$$((\lambda sys / 2) * Ca * MTTR) +$$

$$((\lambda sys / 2) * (1 - Ca) * Cm * (MTTR + TI/2) +$$

$$((\lambda sys / 2) * (1 - Ca) * (1 - Cm) * (Life)$$

Common cause models use the simplex (1oo1) formulas. Systematic failures (λsys) are split fifty-fifty initiating/inhibiting.

In a PLC system, sets of these formulas are used for the CPU assembly and each I/O module type. The systematic failure PFD_{avg} (formulas [g] through [i]) is only calculated once per system. As you can see, the formulas are just simple algebra, but there are so many of them that it becomes unwieldy (to put it mildly) to crunch them manually using a calculator. One can easily automate the process using computer spreadsheets.

When modeling multiple module groups (for example, ten sets of triplicated input modules) it is important to factor the quantities properly. Two possible choices would be as follows:

1. $6 * Qty * (\lambda_d)^2 * (MTTR + (TI / 2))^2$
2. $6 * (Qty * \lambda_d)^2 * (MTTR + (TI / 2))^2$

The difference in results between formula 1 and 2 here can be *two orders of magnitude* in some cases. The first case correctly states that a failure occurs only when two modules *within one set* fails. The second case states that a failure occurs if a module "a" and a module "b" *anywhere in the system fails*. This is usually *not* the case with most systems.

8.7.3   Impact due to Manual Test Interval

The manual test interval can have a significant impact on the overall performance of a safety system. During the test, a simplex system must be off line (to prevent an actual trip due to testing), a dual system is reduced to simplex, and a triplicated system is reduced to dual. The following formulas (for PFD_{avg}) were developed in Geoff R. Duke, "Calculation of Optimum Proof Test Intervals for Maximum Availability":[7]

1oo1   MTD / MTI

1oo2 $2 * MTD * \lambda_d * (((MTI / 2) + MTTR) / MTI)$

2oo2 $2 * (MTD / MTI)$

2oo3 $6 * MTD * \lambda_d * (((MTI / 2) + MTTR) / MTI)$

where:

MTD   =   manual test duration

MTI   =   manual test interval

The PFD_{avg} due to the manual test interval is simply added to the results of the previous PFD_{avg} calculations.

8.8   Analysis of a Relay System

First, we must assume a failure rate (or MTBF) of a relay. Data books will show a considerable range of data for different types of relays. We can assume a one-hundred-year MTBF for an industrial relay. First, we must convert the one-hundred-year MTBF to a failure rate expressed in failures per hour:

$\lambda = 1 / (100 \text{ years/failure} * 8760 \text{ hours/year}) = 1.14 \text{ E-6 failures/hour}$

Next, we must consider how many relays to include in the calculation. Let us assume there will be one relay for each input and output in the system. Let us assume a relatively small interlock group, just eight inputs and two outputs. The system would suffer a nuisance trip if any of the ten relays were to fail open circuit. Assuming a relay is 98 percent fail-safe:

MTBFsp   = $1 / ((1.14 \text{ E-6}) * 0.98 * 10)$
(0.98 represents the failure mode split; 10 represents the qty of relays)

            = $(1 / 1.12 \text{ E-5}) / 8760$ *(need to convert to years)*

            =   10.2 years   *(only use two significant digits: 10)*

For the PFD_{avg} calculation, we need to break the I/O down further. When there is a shutdown demand placed on the system, it comes in on one input only (all eight inputs do not go into alarm at the same time). Therefore, we should only include one input and both outputs in the model; in other words, just three relays:

$$PFD_{avg} = (1.14 \text{ E-6}) * 0.02 * 3 * ((8760 / 2) + 4)$$
(0.02 represents the failure mode split; 3 represents the qty of relays)

$$= 3.00 \text{ E-4}$$

$RRF = 1/3.00 \text{ E-4} = 3,300$*(risk reduction factor = $1/PFD_{avg}$)*

$SA = 1 - (3.00 \text{ E-4}) = .9997 = 99.97\%$*(safety availability = $1 - PFD_{avg}$)*

Referring back to Table 6-4, this represents SIL 3 performance, *but* we have *not* yet included field devices. (We will soon.)

8.9 Analysis of a Solid-state System

Modeling a solid-state system is a useful example because one can clearly see the impact different assumptions have on the model's results. One first needs to assume an MTBF or failure rate for the modules. A reasonable number would be three hundred years MTBF for each failure mode. (Remember that MTBF and life are not the same.) Most all of the modules (input, logic, output, etc.) would have similar numbers.

Let us assume each input and output has its own dedicated module and that two logic (AND / OR) modules are used. Converting three-hundred-year MTBF to failure rate expressed in failures per hour, we get

$\lambda = 1 / (300 \text{ years / failure} * 8760 \text{ hours / year}) = 3.81 \text{ E-7 failures / hour}$

$$MTBFsp = 1 / ((3.81 \text{ E-7}) * 12)$$
(12 represents the qty of modules: 8 input, 2 logic, 2 output)

$$= (1 / (4.57 \text{ E-6})) / 8760 \text{ (need to convert to years)}$$

$$= 24.9 \text{ years} \quad \text{(only use two significant digits: 25)}$$

Some solid-state systems may be supplied with an auto-test unit. Let us first assume the user of the system does not wish to spend the additional money for such a unit. He or she will accept testing the system manually, once per year.

For the PFD_{avg}, RRF, and SA (safety availability) calculation, let us assume the demand comes in on one channel and that one logic module and both output modules must function:

$$PFD_{avg} = (3.81 \text{ E-7}) * 4 * ((8760 / 2) + 4)$$
(4 represents the qty of modules; 1 input, 1 logic, 2 output)

$$= 6.68 \text{ E-3}$$

$$RRF = 1 / 6.68 \text{ E-3} = 150$$

$$SA = 1 - (6.68 \text{ E-4}) = .9933 = 99.33\%$$

Referring back to Table 6-4, this represents SIL 2 (one order of magnitude *less safe* than the relays). We still have not included field devices either. (We will soon.) The reason the performance is relatively poor is because a conventional solid-state system is not fail-safe (like relays), and we did not account for the option of automatic testing. Automatic testing would cycle the system through the test procedure at a frequency not possible manually. In addition, at least one vendor can do this automatic testing on line without removing the system from service. If one assumes 100 percent automatic diagnostic coverage of the auto-test unit, what impact would this have on the results?:

$$PFD_{avg} = (3.81 \text{ E-7}) * 4 * ((0.5 / 2) + 4)$$

$$= 6.48 \text{ E-6}$$

$$RRF = 1 / 6.48 \text{ E-6} = 154,000$$

$$SA = 1 - (6.48 \text{ E-4}) = .999994 = 99.9994\%$$

This represents performance numbers *exceeding* SIL 3 requirements; however, it is based on a faulty assumption (100 percent diagnostic coverage). This may not be a valid assumption with certain systems. (In fact, 100 percent diagnostics is *never* a valid assumption.) Let us assume the system has just 99 percent coverage. This means 99 percent of the failures will have a thirty-minute test interval, while the remaining 1 percent will have a one-year test interval. This simply means we have to do two PFD calculations and add them together. Let us see how much of an impact just 1 percent makes:

$$PFD_{avg} = [(3.81 \text{ E-7}) * 0.99 * 4 * ((0.5 / 2) + 4)]$$
$$+ [(3.81 \text{ E-7}) * 0.01 * 4 * ((8760 / 2) + 4)]$$

$$= 6.41 \text{ E-6} + 6.68 \text{ E-5}$$

$$= 7.32 \text{ E-5}$$

$$RRF = 1 / 7.32 \text{ E-5} = 13,700$$

$$SA = 1 - (7.32 \text{ E-5}) = .99993 = 99.993\%$$

Would you have thought that only a 1 percent change in an assumption alters the final answer one order of magnitude, by a factor of *ten*?

Thus, this system offers a nuisance trip rate similar to that of relays (even though the original failure rate might be viewed by some as pessimistic) and better safety performance, assuming the system is properly tested.

8.10 Analysis of a Simplex PLC System

To be consistent with our earlier modeling examples, we will consider a PLC system with one input and one output module. (Do not attempt to split the calculation down to individual channels of a module, as there are usually common components on a board that could make the entire module fail.) Let us assume the following:

CPU MTBF = 10 years

I/O module MTBF = 50 years

CPU safe failure mode split = 60%

I/O module safe failure mode split = 75%

We can also assume a dual redundant power supply. The probability of a simultaneous failure of two supplies would be orders of magnitude below the simplex components, so, in effect, it may be ignored in the model.

Converting ten- and fifty-year MTBFs to failure rates, we get:

$$\text{CPU } \lambda = 1/(10 \text{ years/failure} * 8760 \text{ hours/year})$$
$$= 1.14 \text{ E-5 failures/hour}$$

$$\text{I/O module } \lambda = 1/(50 \text{ years/failure} * 8760 \text{ hours/year})$$
$$= 2.28 \text{ E-6 failures/hour}$$

$$\text{MTBFsp} = 1 / (((1.14 \text{ E-5}) * 0.6) + ((2.28 \text{ E-6}) * 0.75 * 2))$$
(0.6 and 0.75 represent failure mode splits; 2 represents the quantity of I/O modules)

$$= (1/(6.84 \text{ E-6} + 3.42 \text{ E-6}))/8760 \text{(need to convert to years)}$$

$$= 11 \text{ years}$$

Let us now assume 90 percent diagnostic coverage of the CPU and 50 percent for the I/O. We will need to do four "sub" calculations: CPU covered, CPU uncovered, I/O covered, and I/O uncovered. The automatic test interval would be in seconds or minutes. This is so much less than the four-hour repair time that it can be left out of two of the four calculations.

In terms of a one-year test interval, a four-hour repair time is insignificant, so it too can be left out of the other two calculations. The following *assumes* that the PLC is actually manually tested, something that historically has never even happened with some systems (but obviously should):

$$PFD_{avg} \quad = \quad [(1.14 \text{ E-5}) * 0.9 * 0.4 * 4]$$

(CPU failure rate, diagnostic coverage, failure mode split, repair time)

$$+ [(1.14 \text{ E-5}) * 0.1 * 0.4 * (8760 / 2)]$$

(CPU failure rate, nondiagnostic coverage, failure mode split, test interval)

$$+ [(2.28 \text{ E-6}) * 2 * 0.5 * 0.25 * 4]$$

(I/O module failure rate, quantity, diagnostic coverage, failure mode split, repair time)

$$+ [(2.28 \text{ E-6}) * 2 * 0.5 * 0.25 * (8760 / 2)]$$

(I/O module failure rate, quantity, nondiagnostic coverage, failure mode split, test interval)

$$= \quad (1.64 \text{ E-5}) + (2.00 \text{ E-3}) + (2.28 \text{ E-6}) + (2.50 \text{ E-3})$$

$$PFD_{avg} \quad = \quad 4.5 \text{ E-3}$$

$$RRF \quad = \quad 1 / (4.5 \text{ E-3}) = 220$$

$$SA \quad = \quad 1 - (4.5 \text{ E-3}) = .9955 = 99.55\%$$

So, the nuisance trip rate is about the same as the relays, but the system is *less safe* than relays by one order of magnitude. (This tends to surprise a lot of people.) Remember that we were assuming the PLC was actually checked manually once per year, which is an overly optimistic assumption for some.

8.11 Analysis of a TMR System

The MTBFs and failure mode splits assumed for a PLC are reasonable to use even for a TMR system. The hardware is essentially the same; there's just more of it, with greater diagnostic capabilities. As with the PLC, TMR systems have redundant power supplies, which can initially be neglected in the model.

For a TMR system, let's lump the input module, CPU, and output module together as one "leg" of a triplicated system. This would mean if input module no. 1 and CPU no. 2 both failed, the system would fail. Some TMR systems do in fact operate this way.

Converting ten- and fifty-year MTBFs to failure rates and adding them, we get:

CPU safe λ = 1 / (10 * 8760) = 1.14 E-5 * 0.6 = 6.84 E-6

I/O module safe λ = 1 / (50 * 8760) = 2.28 E-6 * 0.75 = 1.71 E-6

Safe λ of one "leg" of a TMR system = 6.84 E-6 + (2 * 1.71 E-6)
= 1.03 E-5

$$\text{MTBF}_{sp} = (1 / (6 * (1.03 \text{ E-5})^2 * 4)) / 8760 \text{ (need to convert to years)}$$
= 45,000 years

Let's initially assume 99 percent diagnostic coverage of both the CPU and I/O modules. We can therefore group the CPU and I/O modules together and only do two "sub" calculations, instead of the four done with the PLC: CPU + I/O covered, CPU + I/O uncovered. As with the PLC, the automatic test interval would be in seconds or minutes. This is so much less than the four-hour repair time that it can be left out of a portion of the model. In terms of a one-year test interval, a four-hour repair time is insignificant, so it too can be left out of a portion of the model. The following assumes the TMR system is actually manually tested, something that also may never happen in reality with some systems. Converting ten- and fifty-year MTBFs to failure rates and adding them, we get:

CPU dangerous λ = 1 / (10 * 8760) = 1.14 E-5 * 0.4 = 4.56 E-6

I/O module dangerous λ = 1 / (50 * 8760) = 2.28 E-6 * 0.25 = 5.70 E-7

Dangerous λ of one "leg" of a TMR system = 4.56 E-6 + (2 * 5.70 E-7)
= 5.70 E-6

PFD_{avg} = $[((5.70 \text{ E-6} * 0.99)^2 * 4^2]$
(failure rate of one leg, diagnostic coverage, repair time)

+ $[((5.70 \text{ E-6} * 0.01)^2 * (8760 / 2)^2]$
(failure rate of one leg, nondiagnostic coverage, manual test interval)

= (5.09 E-10) + (9.97 E-7)

PFD_{avg} = 9.97 E-7

RRF = 1 / (9.97 E-7) = 1,000,000

SA = 1 - (9.97 E-7) = .999999 = 99.9999%

Based upon these simple assumptions, it would appear that a TMR logic system meets SIL 4 performance requirements, even though we have not yet included field devices.

8.11.1 Common Cause

Both the nuisance trip rate and the RRF in the previous example would appear to be orders-of-magnitude better than the relay and/or solid-state systems. *However*, the input assumptions might be considered a bit unrealistic as there was no accounting for common cause problems. Let us examine how common cause can be included and what sort of impact it will have on the overall performance of a system.

As we have seen, one method for working with common cause is called the beta factor. Essentially, one takes a certain percentage of the overall component failure rate in each failure mode and models it using the simplex formula. Doing this for the MTBFsp and assuming 1 percent common cause, we get:

$$\text{MTBFsp} \quad = \quad (1 \: / \: ((1.03 \: \text{E-5}) * 0.01)) \: / \: 8760 \; \textit{(need to convert to years)}$$

$$= \quad 1{,}100 \text{ years}$$

In other words, our initial estimate of forty-five thousand years is optimistic. Just a 1 percent common cause factor lowers performance more than one order of magnitude and dominates overall system performance. (Similar calculations could be done for the PFD_{avg}, RRF, and SA.)

Another way of accounting for common cause is referred to as systematic or functional failures, first introduced in Chapter 1. Such failures could be an operating system problem, application programming error, maintenance error, periodic lightning strike, and so on. Although it may be very difficult to determine exactly which number to use in a model, it is still easy to account for—one simply uses the exact simplex formulas as before. In redundant systems, performance is usually impacted by another order of magnitude.

8.12 Field Devices

All of the modeling examples given so far were for the logic boxes only, yet the performance requirements stated in the standards are meant to include the field devices (sensors and valves). What impact, if any, will this have on the calculations?

Let's consider the same system as before, one with eight inputs (sensors) and two outputs (valves). Let's simplify and further assume that the sensor and valve have identical MTBFs, one hundred years for both failure modes. (Out of one hundred devices, in one year one causes a nuisance trip, and, when tested yearly, one is found not to be functioning properly.) Since any of the sensors could cause a nuisance trip, all eight should be included in the model. If either valve closing would also result in a nuisance trip, then both valves should be included. This brings the total number of field devices to ten. Calculating the system MTBFsp is simply a matter of dividing the MTBF of a single device by ten:

$$\text{MTBFsp} \quad = \quad 100 \text{ years} / 10$$

$$= \quad 10 \text{ years}$$

If we consider our TMR system along with these ten field devices, the overall MTBFsp number would be as follows:

$$\text{MTBFsp} \quad = \quad 1 / (1 / 1,100 + 1 / 10)$$

$$= \quad 9.9 \text{ years}$$

In other words, as far as nuisance trips are concerned, the TMR logic box is not the problem; the field devices are. What about the safety performance?

Assuming the same MTBF in the inhibiting mode, a discrete switch and simple solenoid and valve (devices with no forms of self-diagnostics), a yearly test interval, and a four-hour repair time and including only one sensor and both valves in the model, the PFD_{avg} may be calculated as follows:

$$\text{PFD}_{avg} \quad = \quad (1.14 \text{ E-6}) * 3 * ((8760 / 2) + 4)$$

$$= \quad 1.50 \text{ E-2}$$

$$\text{RRF} \quad = \quad 1 / 1.50 \text{ E-2} = 67$$

$$\text{SA} \quad = \quad 1 - (1.50 \text{ E-2}) = .985 = 98.5\%$$

If we consider our TMR system along with these three field devices, the overall RRF number would be as follows:

$$\text{RRF} \quad = \quad 1 / (1 / 1,000,000 + 1 / 67)$$

$$= \quad 67$$

Ouch! In other words, our TMR logic box, which by itself could meet SIL 3 requirements, only meets SIL 1 when combined with a small simplex configuration of field devices! Again, the TMR box is not the source of the problem; it's the field devices.

Some might conclude that TMR systems are not required, that one must focus on the field devices. This would not be entirely correct. "Don't throw the baby out with the bathwater." Another way of putting this is that a chain is only as strong as its weakest link. To meet the requirements for the upper safety integrity levels (SILs), one requires either field devices with self-diagnostics and/or redundant field devices, which can all be modeled using the formulas given earlier.

One might also conclude that one does not need a dual or triplicated logic system for SIL 1 or 2 applications. This also would not be entirely correct. Reliability predictions are a useful tool, but they should not be the only deciding factor, for not everything can be quantified. If you were in the market for a new car, would you buy one sight unseen after just looking at a one-page specification sheet without even so much as a picture? There are intangible factors that cannot be expressed numerically. Similarly, certain safety systems have benefits that make them ideally suited for safety, even in SIL 1 applications.

8.13 Engineering Tools Available for Analyzing System Performance

It is useful to first model systems by hand so one realizes that (1) there is no magic involved and (2) you do not need a Ph.D. in math to do it. Crunching numbers by hand, however, can quickly become tedious and boring. Engineering tools are available, however, to automate the modeling process.

One method is to take the formulas presented earlier and build them into a computer spreadsheet. Along with calculating system performance, one can generate charts and diagrams showing the impact of test intervals, diagnostics, and so on. This form of automation greatly helps you understand which factors have a significant impact on system performance and which do not.

A number of commercial programs for performing reliability block diagrams, fault trees, and Markov modeling are available.[8] These programs are generic, however, and are not specific to the unique nature and design of most of the safety systems currently in use.

Control and safety system vendors have developed their own modeling programs. These programs may or may not be made available for users. At least one commercial program specifically designed to model the performance of control and safety systems is available.[9]

Summary

Things are *not* as intuitively obvious as they may seem. Dual is *not* always better than simplex, and triple is *not* always better than dual. Which technology should you use, with what level of redundancy and at what manual test interval? Moreover, what about the field devices? If answers to these questions were easy, it would not have taken the standards groups ten years to resolve these issues, and there would be no need for this book.

We do not design nuclear power plants or aircraft by "gut feeling" or intuition. As engineers, we must rely on quantitative evaluations as the basis for our judgments. Quantitative analysis may be imprecise and imperfect, but it nevertheless is a valuable exercise for the following reasons:

- It provides an early indication of a system's potential to meet the design requirements.

- It enables one to determine the weak link in the system (and fix it, if necessary).

To predict the performance of a system, one needs performance data on all the components. Information is available from user records, vendor records, military-style predictions, and commercially available databases in different industries.

When modeling the performance of an SIS, one needs to consider two failure modes. Safe failures result in nuisance trips and lost production. The preferred term in this mode for system performance is the *nuisance trip rate* (measured in years). Dangerous failures cause problems where the system will not respond when required. Common terms used to quantify performance in this mode are *PFD* (probability of failure on demand), *RRF* (risk reduction factor), and *SA* (safety availability).

A number of modeling methods may be used to predict safety system performance. The ISA technical report TR84.0.02 provides an overview on using simplified equations (to Markov models), fault trees, and full Markov models.[10] Each method has its pros and cons. No method is more "right" or "wrong" than any other, as they are all simplifications and all account for different factors. (In fact, modeling the same system in TR84.0.02 using all three methods produced the same answers.) Using such techniques, one can model different technologies, levels or redundancies, test intervals, and field device configurations. One can model systems using a hand calculator or develop spreadsheets or stand-alone programs to automate and simplify the task.

References

1. Nancy G. Leveson, *Safeware: System Safety and Computers* (Reading, MA: Addison-Wesley, 1995), ISBN 0-201-11972-2.

2. Trevor Kletz, *Computer Control and Human Error* (Houston, TX: Gulf Publishing, 1995), ISBN 0-88415-269-3.

3. OREDA-92 (Offshore Reliability Data), DNV Industry, Veritasveien 1, N-1322 (Høvik, Norway: 1992); American Institute of Chemical Engineers, *Guidelines for Process Equipment Reliability Data with Data Tables* (New York: AIChE, Center for Chemical Process Safety, 1989), ISBN 0-8169-0422-7; and Institute of Electrical and Electronics Engineers, *Equipment Reliability Data for Nuclear-Power Generating Stations,* standard 500-1984 (Piscataway, NJ: IEEE and John Wiley & Sons, 1984), ISBN 471-80785-0.

4. MIL-HDBK 217F.

5. David J. Smith, *Reliability, Maintainability and Risk (Practical Methods for Engineers)*, 4th ed. (London: Butterworth-Heinemann, 1993), ISBN 0-7506-0854-4.

6. William M. Goble, *Evaluating Control System Safety and Reliability: Techniques and Applications* (Research Triangle Park, NC: ISA, 1998).

7. Geoff R. Duke, "Calculation of Optimum Proof Test Intervals for Maximum Availability," *Quality and Reliability Engineering International* Volume2, 1986): 153-58

8. Item Software, Irvine, CA, 1998.

9. CaSSPack (control and safety system modeling package), L&M Engineering, Kingwood, TX, 1998.

10. ISA, *ISA Technical Report* TR84.0.02 (Research Triangle Park, NC: ISA, 1998).

11. P. G. Neumann, *Computer-Related Risks* (Reading, MA: Addison-Wesley, 1995), ISBN 0-201-55805-X.

ISSUES RELATING TO FIELD DEVICES

"When purchasing real estate, the three most important selection criteria are location, location, and location. When purchasing a safety instrumented system, the three most important selection criteria are diagnostics, diagnostics, and diagnostics."

9.1   Introduction

Field devices include sensors, final control elements, field wiring, and other devices hardwired to the input/output terminals of the logic system. These devices are the most critical and probably the most misunderstood and misapplied elements of safety systems.

It is well understood that safety instrumented systems (SIS) include all elements from the sensor to the final elements, including inputs, outputs, power supplies, and the logic solver. In spite of this, the emphasis field devices are paid in the design and application of safety systems is disproportionally low compared to the potential impact that they have on the safety system's overall performance. It is estimated that approximately 90 percent of safety system problems can be associated with field devices. (See the analysis in Section 9.2.)

In the past, relay, solid-state, and conventional programmable logic controllers (PLCs) were used as the logic solvers. PLCs have the advantage of programmable logic, but their use has created issues around the logic system because of the high probability that they will create dangerous failure modes. Hence, the focus on the logic solver has continued. Since economical and "safe PLCs" are now available, the focus is turning toward field devices.

This chapter only briefly covers the types of field devices or their applications that are required for safety systems. Numerous textbooks and publications are available that describe the specific types of field devices required for various applications. The main objective of this chapter is to discuss the issues that have to be addressed in the application of field devices in safety systems.

Some of these issues are as follows:

- diagnostics
- use of transmitters versus switches
- smart transmitters
- smart valves
- use of different technologies
- redundancy
- inferential measurements
- specific application requirements

9.2 Importance of Field Devices

For basic process control systems (BPCS) in process industries, more hardware faults occur in the peripheral equipment—that is, the measuring instruments, transmitters, and the control valves—than in the control room equipment itself. This also applies to safety systems.

9.2.1 Impact of Field Devices on Safety System Integrity Levels

To demonstrate the effect of field devices on the integrity of safety systems, consider the following example:

> A safety instrumented system consisting of a pressure sensor, relay logic, a solenoid valve, and a shutdown valve is proposed to be installed for an SIL 1 application.
>
> Based on the above SIL requirement, the PFD_{avg} should be between 0.1 and 0.01.
>
> The test interval is initially set to be once per year, i.e., TI = 1
>
> The PFD_{avg} for each element of the system can be calculated using the formula

$$PFD_{avg} = \frac{1}{2} \lambda_d * TI$$

where:

λ_d = Fail-to-danger rate

TI = Test frequency or interval

Equipment reliability and performance data for the proposed SIS is given in Table 9-1.

Equipment Item	Fail-to-Danger Rate λ_d (per year)	PFD_{avg}	PDF_{avg} % Contribution
Sensor	0.05	0.025	42
Logic system (4 relays)	0.01	0.005	8
Solenoid valve	0.03	0.015	25
S.D. valve	0.03	0.015	25
TOTAL	0.12	0.06	100

Table 9-1   Reliability and Performance Data

The impact of the field devices on the overall system is 92 percent.

9.2.2   Breakdown of Failures

Figure 9-1 shows the breakdown of failures between the major elements of safety systems based on the data in Table 9-1.

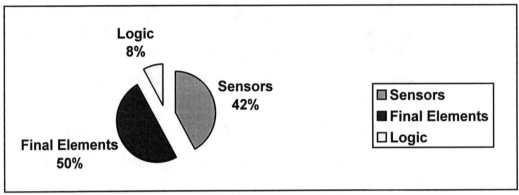

Figure 9-1   SIS—Breakdown of Failures

As a general guide, field devices account for approximately 90 percent of failures, while the logic system accounts for only 10 percent.

The data in Figure 9-1 relates to hardware failures only. External factors also contribute to component failures, for example, poor maintenance procedures, calibration errors, and training. When one takes these external factors into consideration a higher failure rate for the sensors is usually encountered because more activities are usually centered on these devices.

9.3   Sensors

General issues relating to transmitters, sensor diagnostics and smart transmitters are discussed below.

9.3.1   General

Sensors are used to measure process variables, such as temperature, pressure, flow, level, and the like. They may consist of simple pneumatic or electric switches, which change state when a set point is reached, or they may contain pneumatic or electronic analog transmitters, which give a variable output in relation to the strength or level of the process variable.

Sensors, like any other device, may fail in a number of different ways. They may cause nuisance or spurious trips, that is, respond without any corresponding change of input signal. They may also fail dangerously, that is, fail to respond correctly to an actual change of input conditions.

These are the two failure modes of most concern for safety systems.

Specific failure mode examples are as follows:

Switches

- impulse line blockage
- leaking impulse lines
- mechanical damage or internal leakage
- fouling of switch
- corrosion of contacts

Thermocouples

- thermocouple burnout

Transmitters

- fail high or fail low
- smart transmitter left in "forced output" mode
- buildup of liquids in leg lines
- erratic output
- plugged or leaking leg lines
- frozen signal
- drift

Most SIS are designed to be fail-safe. This usually means that when power is lost the safety system makes the process revert to a safe state, which usually means stopping production. You must consider how the sensors should respond in order to be fail-safe. Nuisance trips should be avoided for safety reasons as well. Start-up and shutdown modes of operation involve the highest levels of risk. Also, a system that encounters too many nuisance trips will sooner or later be bypassed, either totally or partially. The general requirements for fail-safe operation are as follows:

- During normal process operation sensor contacts will be closed and energized.

- Transmitter signals would go to a trip state upon failure.

- The output contacts from the logic system will normally be closed.

- The final trip valves would go to a safe position upon air failure.

- Other final devices, for example, motor or turbine, would stop.

Smart transmitters in general can be configured to fail upscale or down-scale in the event of a failure of the internal electronics. Thought should be given to the failure mode for each type of transmitter. An overall, across-the-board recommendation for all transmitters cannot be made. (As with control valves, the failure mode for smart transmitters must be defined. This should be included in the safety requirements specification [SRS]; see Chapter 5.)

Some measurements may be inferred from other variables. For example, if a system is designed to shut down due to high flow, it may be more effec-tive to monitor temperature (if, due to the process, an elevated temperature might also imply a high flow). Inferential measurements can be used to initiate shutdowns. The following should be taken into consid-eration when using such measurements:

- **Response time:** The response of the measured variable may be very slow, that is, changes in temperature versus flow.

- **Correlation:** The relationship between the two variables must be well understood, that is, linear or nonlinear.

- **Accuracy:** The relationship may not provide an accurate repre-sentation.

If uncertainties exist when using the inferred variable, you should con-sider using the measurement for alarm. Inferred variables can also be used as extended diagnostics tools. A comparison between a transmitter or switch output and its inferred value or status can be used to trigger shut-downs or alarms.

Special care should be taken when operating sensors at the low end of their ranges because of potential low-accuracy problems. For example, a sensor designed to operate at 1,000 psi may not be able to differentiate between 20 and 25 psi. Most primary elements have their accuracies stated in percentage of full scale. A 1 percent accuracy transmitter may have an error of 100 percent of measured value while measuring into the first 1 percent of its scale.

The sharing of sensors with the BPCS in general is not recommended for the following reasons:

- The SIS is usually required to operate if a component of the BPCS fails. The failure of the BPCS can be a sensor, part of the PES, or a final element. If a sensor in the BPCS has failed, creating potential hazards, then trying to operate the safety system from the same sensor that has failed doesn't make much sense.

- The procedures in place for the testing, calibration, and mainte-nance of the safety system are usually more stringent than those for the BPCS. The integrity of the safety system will be compromised if changes are made to its sensors without following adequate procedures.

- Online testing of the shared field devices may not be possible.

These rationale also apply to final elements.

9.3.2 Sensor Diagnostics

Improving the diagnostics, and hence the diagnostic coverage, of safety systems improves the safety and reliability of the systems. The following formula shows the impact of diagnostic coverage on the dangerous failure rate of sensors.

$$\lambda^{DD} = \lambda_d C_d$$

$$\lambda^{DU} = \lambda_d(1-C_d)$$

C_d = Dangerous coverage factor

λ_d = Total fail-to-danger rate

λ^{DD} = Fail-to-danger rate - Detected

λ^{DU} = Fail-to-danger rate - Undetected

An increase in the diagnostic coverage factor C_d increases λ^{DD} (the dangerous and detected failure rate) and decreases λ^{DU} (the dangerous and undetected failure rate). Note: Diagnostics do not change the overall failure rate. It is the manner in which the failures are detected that is changed.

One advantage of increasing C_d is that the safety system designers have the option of shutting the process down or only generating an alarm if a dangerous fault is diagnosed with the sensor.

Ideally, we would like to have sensors in which $C_d = 1$, hence $\lambda^{DU} = 0$, and there would be no dangerous undetected failures. Although it is not practical for $C_d = 1$, our objective is to try to get C_d as close to 1 as possible. By improving sensor diagnostics, obviously the need for periodic maintenance and testing will be reduced.

How to improve sensor diagnostics: A very simple, but very effective, method for improving sensor diagnostics is to compare the signal from the trip transmitter with other related variables. Smart transmitters provide diagnostics not normally available with conventional transmitters (see Section 9.3.3 for more details).

Transmitters are now being manufactured and certified specifically for safety applications, and are thus known as "safety transmitters." These devices provide a higher level of robustness, internal redundancy, and diagnostics than conventional and smart transmitters. Rainer Faller's *Safety-Related Field Instruments for the Process Industry* provides additional information on the use and application of these transmitters.[1]

Special external electronic alarm relays can be installed to monitor the performance of the transmitters (see Section 9.5 for additional details).

9.3.3   Smart Transmitters

Some of the advantages of using smart transmitters in safety applications are as follows:

1. better accuracy than conventional transmitters

2. negligible long-term drift and excellent stability

3. improved diagnostics; diagnostic coverage factors is ≈ 50 percent

4. predictable failure modes

5. remote maintenance calibration capabilities

6. less exposure by maintenance technicians to hazardous environments

7. special safe work permits may not be required because of remote maintenance

The main disadvantages are as follows:

1. easy to change parameters (e.g. range) without ensuring that the changes are reflected in the logic and other systems

2. easy to leave the transmitter in a test or "forced output" mode

3. cost—smart transmitters are more expensive than conventional transmitters

4. response may be slower than conventional transmitters

Even with these disadvantages, smart transmitters are increasingly being used for safety applications.

9.4   Final Elements

General issues relating to valves, valve diagnostics and smart valves are discussed below.

9.4.1   General

End devices represent the final stage in implementing the safe shutdown of a process. Final elements or end devices include solenoid valves, trip valves, or motor starters. The more common end devices are solenoid valves, which provide air to a diaphragm or the piston actuator of a process shutdown valve. Usually, removing the air pressure causes the process valve to move to its shutdown position.

Final elements generally have the highest rates of dangerous failure. They are mechanical devices and subject to harsh process conditions. Safety shutoff valves also suffer from the fact that they are usually open and not activated for long periods of time, except for testing. The most common failure modes are when the valve fails to operate after being in a fully open position for an extended period or when the solenoid valve coil has burned out.

Valves should be fail-safe upon a loss of power. This usually requires a spring. Piston-type or hydraulic valves normally require extra large operators with springs or a volume bottle to be fail-safe. The "availability" of the bottles may be inadequate.

Typical failure modes for the final elements are as follows:

Solenoid Valves

- coil burnout
- plugging of ports or vents
- corrosion of terminals or valve body, making solenoid inoperative
- hostile environment leads to sticky valve stem

Trip Valve

- leaking valve
- actuator sizing insufficient to close against shutoff pressure
- sticky valve stem or valve seat
- blocked or crushed air line
- stuck open

There is a tendency to also use the process control valve as the trip valve because it is felt that the control valve is "self-testing." As for sensors, the sharing of final elements with BPCS in general is not recommended (see Section 9.3). Some additional reasons why process control valves should not also be used as trip valves are as follows:

- Control valves account for the largest percentage of instrument failures.

- Process control valves are designed for their control capabilities; leakage, fire safety, and speed of response are some of the parameters that may not be considered when selecting control valves. These are some of the key parameters for trip valves.

Obvious exceptions to these generalizations are motor starters and non-electric drive systems.

In spite of these generalizations process control valves are being used to also provide a trip function whenever it is impractical or not cost-effective to install a separate shutdown valve. For these applications, the solenoid valve should be tubed directly to the actuator, bypassing the valve positioner. Redundant solenoids may need to be used since the valve positioner is now the common weak link hindering the successful operation of the BPCS and safety system. Careful sizing of the solenoid is also required to ensure adequate speed of response of the main valve.

9.4.2   Valve Diagnostics

Trip valves associated with safety systems must operate when a signal is sent from the logic system to the valve. During normal operation these valves are usually fully open and will remain in this state unless tested or activated by the logic system.

The diagnostics associated with valves should then be considered under these two stated conditions:

1. normal operation
2. when activated

1. **Normal operation**: During normal operation the following techniques can be used to diagnose the state of the valve:

   - Online testing (refer to chapter 12)

   - Using a smart valve positioner (refer to section 9.4.3)

   - Indicating an alarm if the valve changes state without a signal from logic.

   - Testing the valve automatically by allowing the logic system to automatically move the valve by approximately 5 percent and monitoring that movement. The logic system sends a signal to the valve for a fixed time (usually a few seconds, based on valve and process dynamics). If the logic system fails to see a movement by means of a position transmitter in the valve within the time an alarm is activated there is a potential problem with the system.

2. **When activated**: Limit switches or position transmitters can be installed to provide feedback to the logic system, indicating that the valve has operated correctly or incorrectly.

9.4.3 Smart Valves

Smart valve positioners are now being installed in trip valves to obtain additional diagnostic data on the health of valves. The positioners can be installed just to provide diagnostic data or they can be used to trip the valve. In the latter case, the trip signal from the logic system is connected to the positioner.

Information that can be obtained from smart valve positioners is as follows:

- position feedback

- problems with the air supply or actuator

- limit switch status

- valve position status

- alarm and self-diagnosis of positioner

9.5 Technologies

Discrete switches do not provide any form of diagnostic information. For example, if under normal circumstances a pressure sensor has a closed set of alarm contacts that are designed to open upon low pressure and the contacts become stuck and cannot open as a result, the sensor has no way to annunciate the failure. The only way to know whether devices such as these are working is to *test* them.

Discrete or on/off switches have been used in the past for the following reasons:

- trips are on/off

- relay logic was widely used in the past

- PLCs were originally designed with no analog inputs

- cost

To address some of these disadvantages, on/off pneumatic indicating controllers with built-in sensors can be used as the primary sensor. The set point of the controller is set at the process trip value. The output signal from the controller is either 3 or 15 psi. This signal is tubed to a pressure switch set at 9 psi, and the contacts from this switch are wired to the logic system.

The advantages of using the on/off controller are as follows:

1. Local indication of process variable.

2. Ease with which sensor can be tested on line—the set point of the controller can be changed to initiate trip.

3. Improved accuracy, rangeability, and reliability.

4. The door to the controller can be key locked for security.

Transmitters, on the other hand, provide an analog signal in relation to the input variable, thus indicating in a limited sense whether the device is functioning properly. Any information is better than none. However, if the transmitter dynamic output is never monitored by the operators or in the logic system, then there really may be no more usable information than a discrete switch. It would be like having a color printer but only printing black-and-white documents. The perceived benefit of having the color printer is illusory if one is unable to use it.

Even though they cost more than switches, analog transmitters are now preferred for the following reasons:

- increased diagnostics

- field indicator is available

- improved reliability

- comparison of signal with BPCS

- single transmitter can replace several switches

- accuracy

Most of the diagnostics presently available in smart transmitters are associated with the protocol inherent in the design of the transmitter, for example, HART (Highway Addressable Remote Transducer). For these smart transmitters it is possible to preset the failure mode of the transmitter, that is, 0 percent, or 100 percent reading. Also, by analyzing the 4-to-20-mA signal from the transmitter it is possible to determine the operational status of the transmitter, for example, <4 mA or >20 mA is an indication that the transmitter reading is faulty.

One device that has recently been introduced by a reputable manufacturer of electronic alarm devices is a HART alarm device. (Devices for other protocols are also available, but this discussion will be restricted to HART.) This device was developed because the manufacturer and users recognized that smart transmitters with HART protocol are capable of providing a wealth of data beyond the simple 4-to-20-mA measurement signal. By connecting the HART alarm device to the 4-to-20-mA loop, the device will provide a transmitter fault signal in addition to the abnormal process condition signal.

The unit can initiate a transmitter fault signal if any of the following occurs:

- a transmitter malfunction
- the sensor input is out of range
- the transmitter analog output is frozen
- the analog output is out of range

By utilizing these diagnostics, the diagnostic coverage of the smart transmitter is estimated to increase from approximately 50 percent to 85 percent.

The two signals from the alarm device (transmitter state and process state) can then be wired to the logic system for either alarm or process shutdown (see Section 9.6, Tables 9-2 and 9-3 for further analysis).

The technology associated with solenoid and trip valves is extremely simple and straightforward, but because of the harsh environment and conditions under which these devices have to operate failures do occur. Better diagnostics and smart valves are now increasingly being utilized to improve the performance of these devices (see Section 9.4).

9.6   Redundancy

If the failure of any one sensor is of concern (i.e., a nuisance trip or a fail-to-function failure), then redundant or multiple sensors may be used. Ideally, the possibility of two sensors failing at the same time should be very remote. Unfortunately, this does not account for common cause failures, which might impact multiple sensors at the same time. Common cause failures are usually associated with human error or external environmental factors such as heat, vibration, corrosion, and plugging. If multiple sensors are used, they should be connected to the process using different taps and separate power supplies so as to avoid common cause plugging and power failures. You might consider using different sensors from different manufacturers or having different maintenance personnel work on the sensors (so as to avoid the possibility that a maintenance technician will incorrectly calibrate all the sensors).

Whenever redundant transmitters are used, an attempt should be made to utilize different technologies, for example, for level sensing, a differential pressure (d/p) cell and a capacitance probe can be considered. It is essential that both sensors have a proven performance history in such applications. If only one sensor has a known proven performance record, then this device should be used for both sensors.

Redundancy should be analyzed for common cause faults.

Diverse redundancy uses a different technology, design, manufacture, software, firmware, and so on to reduce the influence of common cause faults. Diverse redundancy is used if it is required to meet the SIL. Diverse redundancy should not be used where its application can result in the use of lower reliability components that will not meet system reliability requirements. Diverse redundancy can also be accomplished by using different measurements.

9.6.1   Voting Schemes and Redundancy

The redundancy and voting required in the system depends on numerous factors, for example, SIL requirements, spurious trip rate, trip test frequency, and the like. The following table shows the redundant and nonredundant schemes commonly used for sensors and final elements:

SENSORS	
1oo1	Single sensor installed and used if the system meets the performance requirements.
1oo1D	Identical to 1oo1 but additional diagnostics provided above and beyond what is normally available in smart sensor for increased diagnostic coverage. See table 9.2 for details
1oo2	Two sensors installed. Only one required to trip. This scheme is more fail safe than a 1oo1 system, but the nuisance trip rate is doubled.
1oo2D	Identical to 1oo2 but additional diagnostics provided above and beyond what is normally available in smart sensor for increased diagnostic coverage. See table 9.3 for details.
2oo3	Three sensors installed. Two required for trip. Used if the frequency of failures have to be minimized
FINAL ELEMENTS	
1oo1	Single valve installed, and used if the system meets the performance requirements.
1oo2	Two valves installed. Only one required to trip. As for sensors is susceptible to twice the incidence of nuisance trips than 1oo1 systems.
2oo2	Two valves installed. Both are required to trip. Although 2oo2 voting substantially reduces the probability of nuisance trips, it is twice as susceptible to fail-to-danger of undetected faults.

9.6.1.1   Voting Logic for 1oo1D Systems

Figure 9-2 shows how a single smart transmitter with failure diagnostics can be used to reduce the dangerous undetected failure rate λ^{DU}.

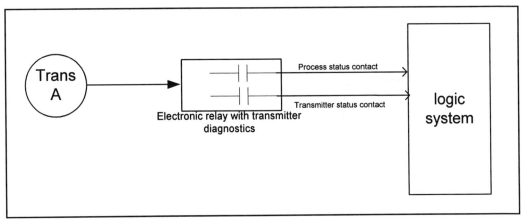

Figure 9-2   **1oo1 Transmitter with Diagnostics**

For this system, transmitter A would be smart with HART protocol. The HART electronic alarm relay monitors the HART digital signal from the transmitter and alarms if either the process variable is outside the set limits or the diagnosis within the HART digital signal detects a failure.

For a single transmitter, Table 9-2 shows the possible conditions.

Condition	Process state as sensed by transmitter 1 = NO TRIP. 0 = TRIP STATE. ? = UNCERTAIN.	Transmitter state sensed by diagnostics 1 = NO ERROR. 0 = FAULT DETECTED.	1oo1D logic status
1	1	1	OK
2	0	1	TRIP
3	?	0	TRIP OR ALARM
Note: An alarm should always be generated if a transmitter fault occurs.			

Table 9-2   **1oo1D Logic Status**

The three conditions shown in Table 9-2 have the following meanings:

Condition 1:   The process state is normal (OK), and no faults have been detected in the transmitter.

Condition 2:   The process state is outside limits, and no faults have been detected in the transmitter. Trip state.

Condition 3:   The process state is unknown since a fault has been detected with the transmitter. (The logic system can be programmed to trip or alarm under these set of conditions. This will impact the diagnostic coverage C, PFD_{avg}, and $MTTF_{sp}$, depending on which selection is made.)

9.6.1.2   Voting Logic for 1oo2D Systems

Figure 9-3 shows the voting for 1oo2D voting using two transmitters A and B.

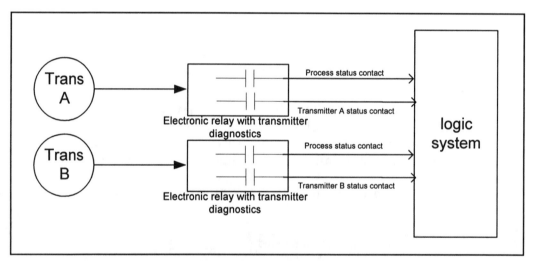

Figure 9-3   1oo2 Transmitters with Diagnostics

The general concept shown in Figure 9-3 is similar to the 1oo1 system. Table 9-3 shows the possible conditions for two transmitters.

For conditions 2 and 3, we have the option of either tripping the process or tripping the alarm if a single transmitter fails. As with the 1oo1D option, this will impact the diagnostic coverage C_d, PFD_{avg}, and $MTTF_{sp}$, depending on which selection is made.

Redundancy and voting is therefore applied to provide enhanced safety integrity and improved fault tolerance. The designer should ensure that the redundancy selected achieves the SIL and reliability requirements for the sensor and final control elements.

Condition	Process state as sensed by transmitter A 1 = NO TRIP. 0 = TRIP STATE. ? = UNCERTAIN.	Transmitter A state sensed by diagnostics 1 = NO ERROR. 0 = FAULT DETECTED.	Process state as sensed by transmitter B 1 = NO TRIP. 0 = TRIP STATE. ? = UNCERTAIN.	Transmitter B state sensed by diagnostics 1 = NO ERROR. 0 = FAULT DETECTED.	1oo2D logic status
1	1	1	1	1	OK
2	?	0	1	1	Alarm or trip
3	1	1	?	0	Alarm or trip
4	0	1	0	1	TRIP
5	?	0	0	1	TRIP
6	0	1	?	0	TRIP
7	?	0	?	0	TRIP
8	0	1	1	1	TRIP
9	1	1	0	1	TRIP
Note: An alarm should always be generated if a transmitter fault occurs.					

Table 9-3   1oo2D Logic Status

An example of this is where the safety system requires 1oo2 architecture, but there is concern about spurious trips. In this situation, the designer may choose a 1oo2D architecture, which may improve reliability without substantially reducing safety integrity, as, for example, in conditions 2 and 3.

9.7   Design Requirements for Field Devices

All field devices must be considered integral parts of the overall safety system and must be selected, designed, and installed to satisfy the system SIL requirements. Due emphasis must always be paid to field devices.

An analysis of the failure modes of the safety instrumented system should be made to ensure that known failures cannot lead to potentially dangerous operating states. Field devices should be completely independent of other field devices associated with the basic process control systems (BPCS) as well as other nonsafety systems, for example, batch or sequence control systems. The sharing of sensors with the BPCS creates maintenance and procedural problems, for example, range change and testing. This should be avoided.

To achieve the required independence, the following components associated with field devices have to be separate from other BPCS devices:

- field sensors, taps, root valves, and impulse lines

- final devices (solenoid valves, main trip valves, etc.)

- cables, junction boxes, air lines, and termination panels associated with the I/O and dedicated panels for the safety system

- power supplies

If the field devices' signals from the safety system have to be sent to the BPCS for data logging or comparison, then the signals between the two systems must be completely isolated, for example, with optical isolators.

The field device installations must conform with applicable local/state codes and federal legislation. Field devices should be designed "fail-safe," that is, the devices must be continually energized or powered so as to maintain continued operation. Loss of power or deactivation initiates a trip.

For noncritical applications, energize-to-trip systems (non-fail-safe) are used to reduce potentially hazardous nuisance trips due to power failures, brownouts, component failures, and open wiring. In such cases, it is possible to install detection systems, for example, pulse testing, to automatically diagnose shorts or opens in the field wiring.

Only proven technology and systems should be used for field devices. A proven and acceptable track record must be established before using field devices for safety systems.

9.7.1   Sensor Design Requirements

The following general requirements have been observed to improve the reliability and performance of sensors:

- Fail-safe systems should use normally closed contacts and normally energized relays and solenoids; that is, during normal process operation the contacts for field switches should be closed for fail-safe operation.

- Sensors should be connected directly to the logic system. They should not be connected to any other system that is not part of the interlock system.

- Smart transmitters are being used more frequently for safety systems because of their enhanced diagnostics and improved reliability. When using such transmitters, procedures should be set up to ensure that they cannot be left in the "forced output" mode, and special procedures must be in place for making configuration/calibration changes to the transmitters.

- Nonsmart analog transmitters tend to have an indeterminate failure mode. Unless it can be demonstrated that a single transmitter installation provides acceptable SIL values, redundant transmitters are recommended (see section 9.6). Separate taps are required to eliminate common mode fail-to-danger faults such as plugged leg lines or the accidental closing of an instrument valve.

- Contacts for electrical switches should be hermetically sealed for greater reliability.

- Electronic trip amplifiers/current relays connected to dedicated safety system analog field transmitters may be used as inputs to the logic system.

- Consider painting all field sensors associated with the safety system with a special color to distinguish them from other devices.

- Where redundant field sensors are used, a discrepancy alarm should be provided to indicate the failure of a single sensor.

- Process connections should be welded, not screwed, for improved reliability.

9.7.1.1   Flow Sensors

The following general guidelines also apply to the most commonly used sensors:

- Vortex or magnetic flowmeters are preferred because of their proven performance and reliability.

- For additional sensor reliability, redundant measuring elements for vortex meters could be considered.

- Head-type flow measurements should be avoided because of their maintenance and reliability problems, for example, leg line pluggage, freeze protection, need for frequent blowdowns, and leakage.

9.7.1.2   Temperature Sensors

Thermocouples (T/C) are commonly used as the source of temperature information for high- or low-temperature trip points. Since the primary failure mode for thermocouples is thermocouple burnout, burnout detection and alarm is mandatory on all T/C inputs. For non-fail-safe systems, upscale burnout indication for low-temperature switches and downscale for high-temperature switches is normally required.

9.7.1.3   Pressure Sensors

Gauge or absolute pressure transmitters are easily installed and extremely reliable. The main precautions when selecting pressure devices are as follows:

- Range selection–Ensure that the trip setting is within 30 percent to 70 percent of the transmitter range.

- Facilities for zeroing transmitters are required.

- Installation should ensure that condensate buildup will not create calibration shift.

- The use of remote diaphragm seals instead of long leg lines is very effective in eliminating "leg line" problems.

9.7.1.4   Level Sensors

Bubbler systems with air or nitrogen purge are very reliable and require very little maintenance.

9.7.2   Final Elements' Design Requirements

Final elements, for example, actuators, should remain in their interlock state after a trip until they are manually reset. They should be reset to only operate if the trip initiators have returned to their normal operating positions.

The following factors should be considered when selecting final devices:

- opening/closing speed
- shutoff differential pressure in both directions of flow
- leakage
- fire resistance
- material suitability/comparability
- diagnostic requirements
- air failure mode, that is, fail open or close
- the need for a position indicator or limit switch
- experience/track record of devices

9.7.2.1   Actuators and Valves

We should avoid installing handjacks on trip valves so the valves will not be left in a jacked or inoperable state. Block and bypass valves should be considered for each actuator valve that fails closed. Block valves alone could be provided for valves that fail open. For cases where blocks and bypass are used, a bleeder valve should be installed upstream of the downstream block valve. Limit switches should be utilized to actuate an automatic bypass alarm when the bypass valve is opened or when a block is closed.

Manual reset solenoid valves should preferably be the push-rod reset type. The lever types can be easily used to defeat the trip system.

Safety instrumented system valves should be dedicated (refer to Section 9.4). The need for and use of nondedicated valves is permitted, but the application must be carefully reviewed during the HAZOP phase.

9.7.2.2   Solenoids

Solenoids are one of the most critical components of final elements. It is important to use a good "industrial-grade" solenoid valve, especially for outside use. The valve must be able to withstand high temperatures, including the heat generated by the coil itself (which is added to the local generated heat), by radiation from furnaces, by exposure to sunbeams, and the like.

Solenoid valves were originally designed to be installed in control room environments. Identical valves with minor modifications for outdoor use are now being used in the field and are required to operate continuously under extreme conditions.

In general, the reliability of solenoids is very low. One of the most common failures is burning out a coil, which causes a false trip. A dual coil would keep the solenoid energized if one coil were to burn out. Solenoids should be tested frequently.

Low-power (24 volts dc and 110 volts ac) solenoids appear to be more reliable because of their low power demand, hence low heat generation. The very-low-power solenoids, such as the intrinsically safe types, are known to be extremely reliable regarding coil burnout, but care has to be taken when choosing them since many are pilot operated. Only "direct-connected," that is, nonpilot, solenoids should be used for safety applications.

9.8   Installation Concerns

Field devices have to be installed so as to satisfy the following requirements:

- environment (temperature, vibration, corrosive gases)
- online testing
- maintenance requirements
- accessibility
- local indicator
- freeze protection

Refer to Chapter 11 for a discussion of installation concerns and details.

9.9   Wiring Field Devices

The common wiring failure modes are

- open/short
- ground fault
- noise-induced voltages

The following are some general guidelines aimed at minimizing wiring problems:

- Fieldbus is not recommended at this time. The Fieldbus Foundation sent a letter to the SP84 committee in 1995 stating that Fieldbus should not be considered for safety systems.

- Each field input and output should have its own dedicated wiring to the logic system.

- It is common practice to sometimes connect several inputs from discrete switches to a single input to the logic system so as to save wiring and input module costs. When this is done with these common connected devices, it is very difficult to troubleshoot and diagnose problems. Connecting a single output to several trip valves is acceptable provided they are all part of the same trip.

- Every field input and output should be fused, or current limited. For PES, this should be done either as a part of the PES system or through the use of external fuses.

- Every field wire should be terminated.

- The system cables, conductors, and junction boxes associated with the safety instrumented system should be segregated from all other control and/or instrumentation wiring and clearly labeled as part of an SIS.

- For relay logic systems the coil circuit should be designed so that the total circuit length does not become excessive. One thousand feet, one way, may be the distance at which you should begin to be concerned that induced voltages via capacitance effect could occur, thus preventing the relay coil from dropping out.

- Voltage segregation should be implemented within systems, junction boxes, and terminal strips and should be clearly labeled to indicate the voltage present. Ground and signal common wires should be clearly identified.

- Individual SIS should utilize dedicated junction boxes or barriers within junction boxes to clearly segregate the wiring of different systems. Segregated terminal blocks in the control room, or some other terminating point for field wiring, should be used. All wiring should be clearly separated on a system-by-system basis.

9.10 Testing Field Devices

Details on testing safety systems are given in Chapter 12.

Summary

Field devices include sensors, final control elements, field wiring, and other devices hardwired to the input/output (I/O) terminals of the logic system. These devices can account for approximately 90 percent of safety system failures.

The issues associated with the application of field devices addressed in this chapter are as follows:

- diagnostics
- use of transmitters versus switches
- smart transmitters
- smart valves
- use of different technologies
- redundancy
- inferential measurements
- specific application requirements

References

1. R. Faller, *Safety-Related Field Instruments for the Process Industry* (Munich: TÜV Product Service GmbH, 1996).

2. ANSI/ISA, *Application of Safety Instrumented Systems for the Process Industries*, ANSI/ISA S84.01 (Research Triangle Park, NC: ISA, 1996), ISBN 1-55617-590-6.

3. American Institute of Chemical Engineers, *Guidelines for Safe Automation of Chemical Processes* (New York: AIChE, Center for Chemical Process Safety, 1993), ISBN 0-8169-0554-1.

4. J. A. Cusimano, *Applying Sensors in Safety Instrumented Systems*, ISA-97 conference proceedings.

10

ENGINEERING A SYSTEM

"Everything should be made as simple as possible, but no simpler."

— Albert Einstein

Designing an SIS involves many choices. Chapter 7 dealt with the pros and cons of various technologies. Choosing the technology for a logic box, however, is not all there is to designing and specifying a system. This chapter will deal with other hardware and management considerations. Many of these aspects of system design are impossible to quantify in a reliability model (the topic of Chapter 8), yet they can have a profound impact on system performance.

The goal of many engineers and companies is to develop a "cookbook" for system design. The more senior engineers, with their experience and knowledge, would write down methods to be followed by the more junior engineers. One must be cautious here, however, for one set of procedures cannot apply to all systems and organizations.

10.1 General Hardware Considerations

The following hardware items should be considered in the design of any system.

10.1.1 Energized versus Deenergized Systems

Most SIS applications are normally energized and deenergize to shutdown. Some applications (typically machine control) are just the opposite—normally deenergized and energize to take action. This difference in system design involves some unique considerations. Systems designed for one may not be (and generally are not) suitable for the other—one cannot just use some form of logic inverter.[1]

For example, in a normally energized system the healthy logic state will be 1, or energized. A revealed failure or the removal of a system component would result in a shutdown (logic 0) of a nonredundant system. For a normally deenergized system the healthy logic state will be 0, or deenergized. A revealed failure or the removal of a system component in such a system should *not* result in a shutdown (logic 1) of a nonredundant system—but it would if the logic were simply inverted. Such systems must be designed differently.

For normally deenergized systems, output solenoid valves should be supplied with latching mechanisms so that when a trip occurs the valve latches in the proper state. Otherwise, if there is a power loss after a shutdown or batteries are drained, the devices could go back to an abnormal running state, which could result in a hazardous condition.

For normally deenergized systems, alarm fuse terminals should be used. These provide alarms in the event that a fuse goes open circuit. Otherwise, an open fuse in a normally deenergized system would not be recognized, thus creating a potentially dangerous failure that could prevent the system from functioning.

For normally deenergized systems, line monitoring (supervising the circuits) of the input and output field cabling is recommended (and required for some systems). Line monitoring the field cabling makes it possible to detect open circuits, which again represent potentially unrecognized failures that could prevent a normally deenergized system from functioning properly.

10.1.2 System Diagnostics

Any SIS, no matter of what technology or level of redundancy, requires thorough diagnostics. Because SISs are by their very nature dormant or passive, certain failures may not be self-revealing. Hidden failures could prevent the system from responding to a true process demand. In the case of "fail-safe" systems (accepting that nothing is 100 percent fail-safe), such as certain relay or European-design solid-state systems, diagnostics may be performed manually (by toggling circuits on and off to see if they respond properly). For most other technologies, additional diagnostic capabilities must be incorporated in the system design.

Some systems are available off the shelf that have very thorough diagnostics. For these systems no extra user interaction is required. For certain other systems, self-diagnostics may be very limited. These systems typically do require extra user interaction.

For example, electronic components may fail on, or energized. In a normally energized system, such a failure would go unnoticed. This is potentially dangerous, for if there were a real process demand the system would be unable to deenergize. General-purpose PLCs typically have few diagnostics in the I/O modules (low diagnostic coverage). Some users have even written papers on this subject and described the extra I/O modules and application programming that were required to give the system an "acceptable" level of diagnostics.

10.1.3 Minimize Common Cause

Common cause failures are single stressors or faults that make redundant systems fail. Obvious examples are power (e.g., simplex power to a triplicated controller), software (a bug will make all channels fail), and backup switches (if the switch does not work the dual redundant system will fail). Unfortunately, not all failures are so obvious. Studies in nuclear power plants have found that approximately 25 percent of all failures in power generating stations were due to common cause problems.[2] Aircraft have crashed because triplicated control lines were severed. A seven-way redundant government computer system failed when a single fiber-optic cable was cut.[3]

A vicious cycle begins to emerge as redundant systems introduce more complexity, which further exacerbates the common cause problem, which can lead to even more equipment being installed. When redundant systems are used, common cause influences tend to be—excluding inadvertent human intervention—external: temperature, vibration, power surges, RFI. Redundancy is effective against random hardware failures but not against design or other systematic errors. Two common methods to reduce such common cause problems are to use either physically separated or diverse components/systems.

10.1.4 Panel Size and Layout

One of the first documents produced during the design stages of a system are *general arrangement* drawings. These drawings identify the overall size and weight of the cabinets. More than one story has been told of how drawings were produced, reviewed, and signed by clients and systems were built and shipped only to find that they were too large to fit between the control room doors.

Consideration should be given to how personnel will access the system. Technicians must have adequate room to access various parts within the cabinet, otherwise someone squeezing between two pieces of equipment might accidentally trip a circuit breaker. Adequate lighting should also be provided inside the cabinet, and parts should be clearly labeled.

10.1.5 Environmental Considerations

It is important to specify the environmental requirements in a system specification. Adequate protection should be provided against temperature, humidity, vibration, electrical noise, grounding, contaminants, and so on. For example, one of the authors knows of one system where the user wanted a software programmable system, but there was no more room in the control building. The panel was to be located outdoors, in South Texas, next to an incinerator, in a Class I, Division II (explosion-proof) area. The system was supplied in a purged, weatherproof enclosure with vortex air coolers to keep it cool in the summer and a heating element to prevent condensation in cold weather. If the integrator was not made aware of the unique location issues, the system would not have operated for very long.

In general, increasing the temperature of electronics by 10°C decreases their life by approximately 50 percent. In order to have a reliable system cabinet ventilation (or fans) should keep the temperature inside the panel well below the manufacturer's specification. Similar consideration must be given to vibration, corrosive gases, and so on.

10.1.6 Power

Power, like many other topics, is not a black-and-white issue. There is no clear-cut "right" versus "wrong" way of handling power. Simply put, there is a need for clean, protected, and regulated power. There should be an isolation transformer to protect against spikes, transients, noise, over- and undervoltage, and the like.

Critical safety systems should have redundant power. There are many ways to achieve this. Each system and/or method will have pros and cons unique to each application. Each system must be evaluated for the intended usage. Will switching to the backup unit truly be "bumpless"? What may work for one installation may not work for another. Backup batteries, like anything else, will not work if they are not maintained. (Can anything truly be "uninterruptible"?!) *All* portions of the system must be periodically tested. Many of these problems can be overcome by using redundant DC power sources. Diode OR-ing them together eliminates AC switches.

Power supplies should be mounted to afford the best heat dissipation while still allowing for easy access and troubleshooting.

10.1.7   Grounding

Proper grounding is also important for a successful installation. Grounding is usually more critical for software/electronic systems than for older electric systems. Manufacturers' recommendations must be closely followed. Grounding bus bars should be mounted on the ceiling or sidewall of a panel, not on the floor. (It's much more difficult to short out a bus bar by dropping a wrench "up.") A partial checklist of items to check for would include corrosion, cathodic protection, static electricity, and intrinsic safety barriers.

10.1.8   Selection of Switches and Relays

Relays should be sized not only for the maximum load (to prevent welding of the contacts—the most dangerous failure in an SIS) but for the minimum load as well. If this is not accounted for, the trickle current in the circuit might not be enough to keep the relay contacts clean, and circuits may "open" years later because of dirty contacts. This can be difficult to diagnose, as it is generally not the first thing one looks for.

10.1.9   Bypasses

Bypasses are required for certain maintenance activities as well as during start-up. Leaving systems in bypass, however, is potentially *very* dangerous. If the process is hazardous while a SIS function is being bypassed, administrative controls and written procedures should be provided to maintain the safety of the process.

Bypasses may be implemented in many different ways. Installing jumper wires is potentially dangerous because there is no external indication that something is actually in bypass. Operators may be wondering why the system is not working properly and have no clue that some technician somewhere else has placed something in bypass for maintenance work.

Virtually all PLCs can use "forcing" of I/O to accomplish the same thing as a bypass. The ease with which this can be done in some systems is a cause for concern. Leaving I/O points forced, without any indication of such a condition, is potentially very dangerous.

Bypass administration is very important to assure that bypasses are promptly and properly removed. (This is similar to hot work permits—a problem that contributed to the Piper Alpha disaster in the North Sea where over 160 people died.) Strict documentation and control are required. Some type of bypass form should be used. This form should have operations' knowledge and approval, time limits should be stated, the form should be renewed on shift change, and it may require different levels of approval.

With many systems, when an input is in bypass there is no indication at the panel of the true state of the input. How do you know whether the signal is healthy or not before you turn the bypass off? What if the input goes into alarm while it is in bypass—how can you tell? It is better to design a bypass so that even when the input is in bypass, the system still provides an indication of the true state of the field device.

Automatically returning to normal conditions after a start-up bypass can be accomplished in some processes. For instance, a flame detector will need to be bypassed in order to start up a burner system. Once a flame is detected, the bypass is disabled because it is no longer required or desired.

Serious accidents have been documented in the chemical industry while systems were temporarily taken out of service or portions were operated manually without the knowledge of others.[3]

10.1.10 Application Software

"And they looked upon the software, and saw that it was good. But they just had to add this one other feature …"

— G. F. McCormick

Reusing existing software or logic diagrams (i.e., programming a PLC solely based upon the ladder logic diagrams of the relay system it is replacing) does not increase safety and may in fact actually *decrease* it.[4] As stated earlier, most errors can be traced back to the requirements specifications. If the requirements were not carefully reviewed during the initial application, faults may still be present. In addition, hazards associated with the new system could not have been considered when the original system was installed (e.g., a PLC is not fail-safe in the same way that relays are).

One of the most common programming languages for software-based systems is ladder logic. Ladder logic is relatively simple and generally makes sense to the person who wrote it (at least for that week). Unfortunately, the person who has to make changes to the system years later will probably not be the same one who originally programmed it. Unusual and "custom" programming tends to be difficult to maintain. This is not to imply that it should never be done, for it is very easy to fully annotate ladder logic programs with detailed comments, although that obviously does not guarantee clarity.

Different languages are available for programming software-based systems. Ladder logic is the most common language because it was the natural outgrowth of hardwired relay-based systems. Ladder logic, however, may not be considered the most appropriate language for some other kinds of functionality. For example, sequential or batch operations are more easily handled by another language—sequential function charts (SFCs). Math functions are more appropriate for structured text. The IEC 61131-3 standard consists of five languages: ladder logic, function blocks, sequential function charts, structured text, and instruction list. Using a system that implements multiple languages allows various functions to be done in the most appropriate language rather than trying to squeeze things into a language that may end up being somewhat awkward.

Some have considered having diverse redundant systems programmed by different teams, using different logic boxes and different software packages. Not only does this become a design, debugging, maintenance, and documentation nightmare, but history has shown it is not effective. All the programs must be written to a common specification, and this is where most of the problems lie. Similar problems exist when testing software; the testers may omit the same unusual cases that the developers overlooked.

10.1.11 Functional Testing

How will the system be tested? Will portions be bypassed and tested manually, or will some form of automatic testing be incorporated, even for the field devices? The functionality of the entire system should be checked before it is placed in service. This means checking the sensors, logic box, output devices, communication links, and other interfaces. It is best to test the entire system as a whole. If the individual portions are tested separately, the potential problem of having certain interfaces between the components not working as expected may arise. All testing procedures and results should be documented.

10.1.12 Security

How secure will the system be? How will access be controlled? For example, should someone be allowed to suck a program out of a PLC?

One of the authors knows one engineer who said he was the only individual, with the only portable PC, responsible for making changes to a certain system. Yet every time he hooked up to it there was a different version of the program running in the controller than he had on his PC. How was this possible? Some systems do not have any real security features. If this is the case, management procedures must be enforced to prevent unauthorized access.

One of the authors heard a story of a disgruntled engineer who through his home computer and modem was able to break into the plant DCS and alter its database. Again, something like this should never be possible with an SIS. The age-old problem, however, is that hindsight is always twenty-twenty. It is a bit more difficult to have foresight and predict events that may never take place. One might want to keep a variation of Murphy's Law in mind: If the system allows it, and someone can do it, at some point they will—no matter how many warnings they may be given not to.

Keys may be used to control access, but management procedures must not allow the keys to be left in place. Passwords may allow various levels of access into systems, but password access must also be strictly controlled (i.e., do not leave the passwords taped to the operator interface).

10.1.13 Operator Interfaces

There are many different devices that may be used as an operator interface, such as CRTs, alarms, lights, push buttons, and the like. All of these may be used to communicate information to operators, such as shutdown action taken, bypass logs, system diagnostics, sensor, logic box, final element status, loss of energy that impacts safety, failure of environmental equipment, and so on. The operation of the system, however, must not be dependent upon the interface because it may not always be functioning or available.

The following questions should be answered in the requirements specifications for every data item to be displayed:[4]

1. What events cause this item to be displayed?

2. Can, and should, the displayed item be updated, and if so, by what events (e.g., time, actions taken, etc.)?

3. What events should cause the data displayed to disappear?

Failure to specify these criteria is a common cause of specification incompleteness for the interface and a potential source of hazards.

It is important to provide certain information to the operators (but not to the point of overload), yet it is just as important to limit what operators are actually able to do at the interface. If a portion of the system is placed in bypass, what indication (if any) is there to alert others?

The interface is important, but it should not be critical or required for proper operation. It should not be possible to modify the operation of the safety system from the operator interface. Thought should be given as to what should happen if the interface screens go blank (which has been reported on more than one occasion).

Warning signals should not be present too frequently or for too long a time because people tend to become insensitive to constant stimuli.

10.2 General Management Considerations

The following management items should be considered in the design of any system.

10.2.1 Job Timing and Definition

Many system problems are caused by two simple factors: a combination of poor definition and timing. For example, the scope of the job may have been too vague, the functional specification may have been incomplete (or nonexistent at the time of the initial purchase or even of the system design), or the job may have been due last week. Chapter 1 summarized the findings of the U.K. Health and Safety Executive that 44 percent of problems were caused by incorrect specifications (functional and integrity).

10.2.2 Personnel

Once a contract is awarded it is important not to change the people in the middle of the job. Although reams of information are no doubt part of every contract, Murphy's Law predicts that the one thing that isn't will crop up to bite you. That one thing exists only in someone's head, and when that person is transferred that piece of information goes with him or her, only to rear its ugly head at the worst possible moment.

No matter how much is put in writing, specifications are always open to interpretation. If the people are changed, the project's direction will be lost while the new people reformulate the prevailing opinion. This may have dire consequences for the delivery and price of a contract, since items may need to be renegotiated.

10.2.3   Communication between Parties

It is important to establish clear lines of communication between the companies involved. The project manager at company A should be the only one talking to the project manager at company B. If other parties start getting involved, certain items will doubtless not be documented and will inevitably fall through cracks, only to cause major headaches later.

10.2.4   Documentation

Will documentation be provided in the user's standard format, the engineering firm's standard format, or the integrator's standard format? What size will the drawings be? Will they be provided on paper only or in digital format? If so, using which program? Will there be 200 loop diagrams or one "typical" along with a table outlining the information on the other 199 loops? Little things like this can easily be neglected until they become major points of contention.

Summary

Designing an SIS involves many choices. This chapter discussed several hardware and management issues that, although difficult to quantify in a reliability model, are vital to the proper and continued long-term operation of the system.

References

1. Paul Gruhn and Allan Rentcome, "Safety Control System Design: Are All the Bases Covered?" *Control* V:7, July 1992, pp. 38-42.

2. David J. Smith, *Reliability, Maintainability, and Risk (Practical Methods for Engineers)*, 4th ed. (London: Butterworth-Heinemann, 1993), ISBN 0-7506-0854-4.

3. P. G. Neumann, *Computer-Related Risks* (Reading, MA: Addison-Wesley, 1995), ISBN 0-201-55805-X.

4. Nancy G. Leveson, *Safeware: System Safety and Computers* (Reading, MA: Addison-Wesley, 1995), ISBN 0-201-11972-2.

11

INSTALLING A SYSTEM

So much for using two separate contractors.

Gruhn

11.1 Introduction

Although the title of this chapter is "Installing a System," we intend to discuss all activities from the completion of design of the safety system to its successful operation. The key activities are as follows:

- factory acceptance test (FAT) of logic system
- programmable electronic systems (PES) logic simulation
- field installation of complete system
- installation checks (wiring, field devices, and logic system I/O)
- pre-start-up acceptance test (PSAT)
- training of operations and support personnel
- handover to operations
- start-up
- post-start-up activities

The overall objective is to ensure that safety systems are installed in accordance with the detail design, that they perform in accordance with the safety requirement specifications (SRS), and that personnel are adequately trained to operate and support the systems. For this to be accomplished effectively, the activities just listed must be performed.

In its examination of over thirty accidents, the English Health and Safety Executive indicated that 6 percent of the accidents were due to "installation and commissioning." This low figure may be somewhat misleading because the potential for major problems is very high if safety systems are not installed as designed or are not commissioned in accordance with the SRS. Some potential problems are as follows:

- Installation errors. Numerous errors could be encountered, ranging from improper or incorrect terminations to the wrong device being installed for a particular service.

- Inadequate testing to capture installation and possible design errors. This normally occurs if the testing is not well planned, the personnel doing the testing are not adequately trained, or the testing procedures and documentation are poor.

- Use of substandard "free-issue" components by the installation contractor. The major components are usually well specified. Minor items, for example, terminal strips and connectors, have to be of acceptable quality to avoid future problems.

- Unapproved or unwarranted changes are made to the system during construction, testing, or start-up. It is very easy and tempting to make quick changes to get a system "operational." Any changes must follow a management-of-change (MOC) procedure.

- Documentation not updated to reflect as-built condition. This usually creates a problem when maintenance work has to be done on the system.

- Inadequate training of operations and support personnel. A major commitment has to be made by management to ensure that all personnel involved with the safety system are trained.

- Temporary bypasses or "forces" not removed after testing or start-up. With programmable systems, it is easy to bypass or force outputs. Jumpers on terminal strips may remain unnoticed for very long periods. This has to be controlled.

11.2 Factory Acceptance Tests (FAT)

Factory acceptance tests (FAT) usually pertain to the logic system and operator's interface, irrespective of the logic technology being used. Even if the logic system is relay logic with twenty relays or a complex programmable electronic system (PES) with hundreds of I/Os, the system should be tested thoroughly before being shipped from the vendor. The FAT is considered part of the installation activity because the individuals involved with the construction and checkout of the system should participate in this activity.

These tests are usually completed at the vendor's site prior to shipment to the end user to test the hardware and software being supplied. The following sections describe the benefits, the participants, what should be included in the FAT, and how the testing is done. More details regarding FATs can be obtained from *Guidelines for Safe Automation of Chemical Processes* from the American Institute of Chemical Engineers

11.2.1   Benefits of FATs

- The system can be reviewed and tested in a controlled and non-stressful environment, away from the plant site.

- Any problems encountered can be easily resolved and corrected because of the resources available at the vendor's site.

- FATs provide excellent training for support personnel.

- The design personnel's understanding of the system increases with FATs, and they have the opportunity to clarify and rectify any misunderstandings.

- All hardware and software can be thoroughly tested.

- The team-building process between the vendor and design and support personnel is strengthened.

11.2.2   Participants in FATs

The number of participants depends on the complexity and size of the system. It is essential that the responsibilities of each individual be well defined. As a minimum, the following should participate:

- Vendor representative(s).

- User design representative. This is usually the individual who has overall design responsibility for the SIS. This individual should have the lead role for the FAT, be responsible for coordinating the FAT, and be responsible for preparing the required testing procedures.

- End-user support personnel.

11.2.3   What Is Tested

- complete logic system hardware, including I/O modules, terminations, internal cabling, processors, communication modules, and operator's interface.
- auto switchover and redundancy
- software (operations and application)

11.2.4   How the System Is Tested

The testing has to be based on a well-documented procedure, which is approved by both the vendor and user. The testing is done by using the following methods:

- visual inspection
- injecting inputs, usually digital, pulse, 4-to-20 mA, or T/C, and observing the system response
- driving outputs, usually digital or 4-to-20 mA
- creating various failure scenarios to test backup systems
- testing noninteractive logic, that is, logic that does not require feedback from field devices to operate

The complete logic system, including the I/O modules, can be wired to switches, 4-to-20-mA potentiometers, and pilot lights, each labeled as a field device. By actuating the switches and potentiometers, the testing crew "simulates" the operation of the unit. By noting the status of the output lights, the testing crew can see the status of each "solenoid" or "motor" reaction and compare this to the logic or cause-and-effect diagrams to verify compliance. Those tests are manual hand-driven signals. It also helps to check for hardware problems, including I/O failures. The logic interaction is not normally tested using this method. To test the logic thoroughly before start-up, some form of logic simulation is usually required. This is described in Section 11.3.

11.3 PES Logic Simulation

Programmable electronic systems (PES) are increasingly being used as the logic solver for safety systems. One of the advantages in using programmable systems is that they make it possible to test the logic in the PES by using the PES processor to perform the simulation. This technique is commonly known as "emulation within the processor."

This section discusses the benefits of PES logic simulation and how such simulation is implemented. The following issues will be reviewed:

1. What is PES logic simulation?
2. Why simulate?
3. When to simulate?
4. Where to simulate?
5. How to simulate?

11.3.1   What Is SIS Logic Simulation?

- It is a means for functionally testing PES logic and the operators' interface without the field I/Os being connected. The testing is accomplished prior to field installation with limited resources away from the plant environment. The operators' interface can be a hardwired panel or a CRT. The interface that is to be used on site should be used in the simulation if possible.

- Since the SIS logic will be completely tested without the physical I/Os being connected, we assume that all field devices are functioning correctly. With the simulation program, the PES will respond as if the I/O were actually connected and functioning.

SIS logic simulation *is not*

- process simulation. The SIS logic is basically discrete, with digital or analog I/O. The simulation is only for the SIS logic and not the rest of the plant. The logic associated with the basic process control system (BPCS) is not included.

- a mathematical model of the process being simulated on a computer. Only the logic in the PES will be tested.

- a simulation package totally independent of SIS logic. It is possible to purchase off-the-shelf simulation packages. These packages have the advantage of easily incorporating the complex dynamics and timing of the process. For SIS, this is not usually an essential requirement. It is simpler to write the simulation within the PES as a completely separate block, independent of the main SIS logic.

11.3.2   Why Simulate?

There are seven basic reasons to simulate the PES logic:

- Cost: engineering and start-up. It is easier and more cost-effective to rectify software or programming problems off site in a controlled environment than in the plant prior to start-up.

- The project schedule can be better maintained because field problems are minimized.

- No surprises at start-up. The logic has been tested, and operations has already "kicked the tires."

- The simulation is an excellent training tool (pre- and post-start-up).

- Early users provide feedback through their interaction with the simulation model.

- Early bug fixes. No system is perfect; bugs will exist. The earlier they are identified and corrected the better.

- Ease of commissioning: software and hardware issues during start-up can be well segregated since there is full confidence that the logic has been fully tested. Less finger-pointing will occur.

11.3.3   When to Simulate?

There at least three situations in which PES logic simulation is beneficial:

- Because no physical I/O is required, the simulation can be done in the designer's office on an ongoing basis as the logic is being developed. Only the main processor is usually required.

- During the FAT, logic simulation serves as a valuable tool for completing the FAT.

- Prior to start-up and before the PSAT, logic simulation allows you to check communications and logic. Usually, the logic has changed between FAT and PSAT.

11.3.4   Where to Simulate?

PES logic simulation may be performed on or off site:

- off site for program development, FAT, and training
- on site prior to PSAT and as an ongoing refresher training tool

11.3.5 How to Simulate?

There are a number of ways to simulate PES logic: Write simulation logic in the PES processor. Outputs are linked to inputs within the simulation, so when an output is sent from the logic system, confirmation that the output device has operated is obtained. In this way, the logic becomes interactive with the field devices. Timing can be included within logic to represent the dynamics of the field devices. The complexity and accuracy of the simulation will depend on the amount of time, effort, and resources spent. This will depend on the requirements and expectations of the end user.

It is essential that the simulation program be completely separate from the main logic. The simulation logic has to be completely removed prior to the pre-start-up acceptance test. For this to be done, the main logic must not be touched. Trained and experienced personnel must write simulation packages.

Utilize the SIS operators' interface as the interface for testing logic and for training. Figure 11-1 shows the setup of the simulation system.

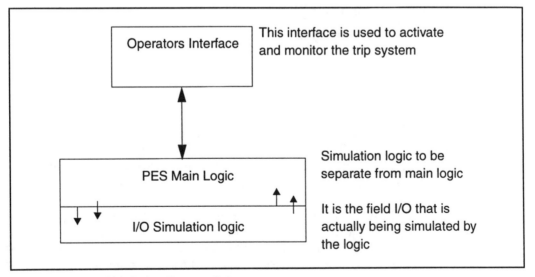

Figure 11-1 Simulation System Setup

To illustrate how the simulation works, let us consider the start/stop logic for a motor. The PES logic for the motor will be as shown in Figure 11-2. The start/stop push buttons can be hardwired push buttons on a panel or part of a CRT interface.

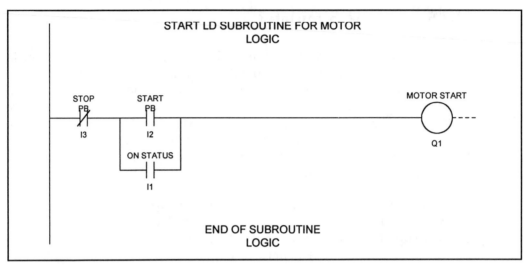

Figure 11-2   Motor Start/Stop Logic

The PES inputs will be:

      I1: Motor run status from motor contactor

      I2: Start input signal

      I3: Stop input signal

The PES output will be:

      Q1:Output to motor starter

If the I/O is not connected to the PES, the logic will not function because the feedback signal (I1) to lock in the start coil will not exist. The objective of the simulation/emulation is to make this happen without actually connecting the field I/O.

Figure 11-3 shows how this can be accomplished in a very simple way, that is, by using the simulation/emulation logic. When Q1 is activated, an internal reference M1 will change state. This value will then be moved to I1 via the bit move block. The PES logic will then respond as if the motor has started.

The motor logic and the simulation logic must be completely separated, that is, written in separate subroutines. The two subroutines will then be linked together in the main program as shown in Figure 11-4. After testing, the simulation subroutine can be removed completely from the PES or disabled.

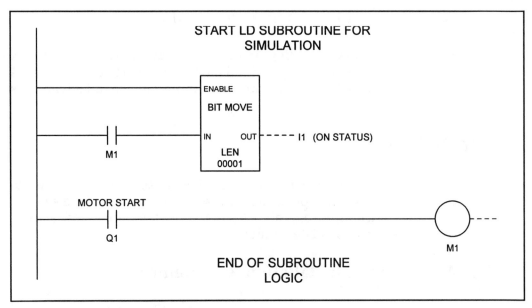

Figure 11-3   Simulation Logic

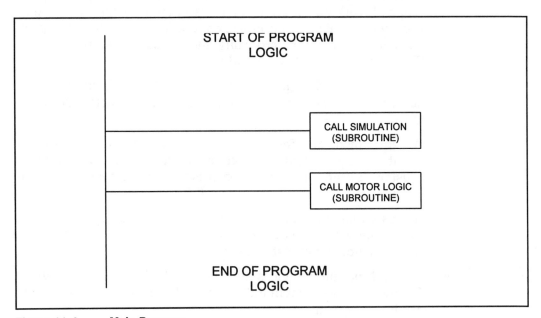

Figure 11-4   Main Program

11.3.6   General Comments on Simulation

Some vendors are now providing simulation capabilities within their PESs to help end users perform logic simulation more effectively. End users must be able to better understand and fully appreciate the full advantages of logic simulation.

11.4 Installation

Installation covers the installation of all equipment associated with the SIS by an installation contractor. It includes sensors, final control elements, field wiring, junction boxes, cabinets, PES, operators' interface, alarms, and other hardware associated with the SIS.

11.4.1   General Installation Requirements

- Usually the installation work for the SIS is part of the overall instrument/electrical installation, and the same contractor/ workforce does the work. Consider splitting or separating the scope of work for the SIS from the rest of the instrument and electrical work. This will minimize common cause installation problems and reinforce the criticality and special requirements of the SIS, for example, in testing and training.

- Ensure that the design package to the contractor is complete and accurate. The training and experience of the contractor are important.

- All equipment and installation must comply with all code and statutory requirements in effect at the local site. The contractor must understand these requirements and ensure compliance with them.

- It is essential that all free-issue material supplied by the contractor for the service be of superior quality. Detailed specs for these items are not always established.

- All field-mounted devices must be installed in a way that allows for easy access so that maintenance and specifically online testing can be carried out.

- No change to the calibration or settings of the field devices must be made by the installation contractor.

- Prior to installation, care must be taken to protect all field devices from physical or environmental damage.

- No changes or deviations from the design drawings are to be made by the contractor without written approval. These changes should be recorded on a set of as-built drawings.

11.5 Installation Checks

Installation checks ensure that the SIS is installed in accordance with the detailed design and is ready for the pre-start-up acceptance test (PSAT). The activities of the test confirm that the equipment and wiring are properly installed and that the field devices are operational/functional.

The installation checks are best completed by breaking the work into two separate and distinct phases, that is:

1. **Device physical checkout.** This is a check of the way in which the field devices are physically installed and of wiring, terminations, tagging, and junction boxes. This phase is usually completed by the installation contractor.

2. **Device functional checkout.** This is the functional check of field devices and logic system I/O after the system is powered up. This phase may be completed by the installation contractor or an independent crew.

11.5.1 Device Physical Checkout

This section contains a generic checkout form that can be modified to suit a particular installation. It is best to have a separate checkout form for each device.

The form shown here will help ensure that the devices are properly installed and that the wiring and termination for transmitters, switches, valves, junction boxes, and logic system I/O are properly checked. The reference drawings used for the physical checkout are usually the wiring or loop diagrams.

Device Physical Checkout Form

> **INSTRUCTIONS:** Use the reference drawing to facilitate in the checking. As items are being checked and are correct, identify with a BLUE LINER. Correct any minor changes such as wiring terminations or wire numbering, on the drawings. Identify on reference drawings any corrections with a RED pencil.

DEVICE NUMBER: _____   **REF. DRAWING NUMBER:** _____

A = Item checked B = Recheck required. C = Re-check completed satisfactorily.	A	B	C	Comments
TRANSMITTER/ DIGITAL DEVICES				
Verify that mechanical mounting of device is correct and adequate				
Verify that transmitter/sensor/switch is correctly tagged.				
Verify that sensor lines are correctly connected to the field device and sensor point.				
Verify that the nameplate data on the field device is as per the device specification sheet.				
Verify that signal wires are correctly terminated and identified at the field device.				
VALVES and SOLENOIDS:				
Verify that the nameplate data on the valve is as per the device specification sheet.				
Verify the valve air supply is correctly connected to the solenoid.				
Verify the air signal line is correctly connected to the solenoid and valve.				
Verify wiring terminations at solenoid are correct.				
GROUNDING:				
Verify that all shield wires are properly grounded according to specifications/wiring diagrams				
Verify that all grounds and shields are properly secured to panels.				
JUNCTION BOX:				
Verify that junction box is correctly tagged.				
Verify that cables to and from JB are correctly identified.				
Verify that input / output signal wires are correctly terminated in junction box.				
Verify that any auxiliary signal wires are correctly terminated in the junction box.				
Verify the color of wire is correct at terminal in the junction box.				
PES OR RELAY PANELS:				
Verify that the panel is correctly tagged.				
Verify wiring is terminated at correct terminals.				
Verify that input and output cables are correctly tagged and properly secured to the panel.				
CONTINUITY TEST:.				
Verify that each device wire and associated equipment wire is rung out from primary point to the final terminal block in each panel.				

DATE CHECKED: _____   **CHECKED BY:** _____

11.5.2   Device Functional Checkout

These checks are intended to confirm the following:

- energy sources are operational
- all instruments have been properly calibrated
- field devices are operational
- logic solver and input/output are operational

These checks should only commence after the physical checks are complete. This section includes a generic checkout form that can be modified to suit the particular installation. It is best to have a separate checkout form for each device.

Device Functional Checkout

INSTRUCTIONS: The functional checkout is to be carried out after the physical checkout has been signed off and accepted. This checkout consists of doing a functionality check on the devices from its point of process origin to the final control element, by introducing simulated signals to verify correct operation. Signals injected to/from field transmitters/switches/final devices to the logic system are verified. Use the referenced drawings to identify, and highlight with a BLUE liner all completions. If check out is not correct or does not pass the acceptance test, identify it in RED.

ENSURE THAT ALL SAFETY PROCEDURES ARE FOLLOWED IN COMPLETING FUNCTIONAL CHECKOUT SINCE DANGEROUS SIGNALS WILL BE APPLIED TO EQUIPMENT.

DEVICE NUMBER: **REFERENCE DWG NUMBER:**

A – Item checked **B** – Recheck required **C** – Recheck completed satisfactorily.		A	B	C	Comments
TRANSMITTERS					
Generate 4-20ma signals at the transmitter by applying a simulated process signal. The simulated signals to be 0%, 25%, 50%, 75% and 100% of the full range.	0%				
	25%				
	50%				
Verify the simulated signals are received at the display area	75%				
	100%				
DISCRETE DEVICES					
Simulate correct contact action at primary device. Verify appropriate logic system response in accordance with the signal.					
FINAL ELEMENTS					
Generate discrete output signal from logic system. Confirm correct switched output at final control device.					

DATE CHECKED:_____   **CHECKED BY:**_____

11.6 Pre-start-up Acceptance Test (PSAT)

These tests are intended to confirm the correct functionality of the *complete safety system*, including the logic. The checkout is done after the device functional checkout has been successfully completed.

ANSI/ISA S84.01-1996 has provided the following comprehensive description for the PSAT:

> *A PSAT provides a full functional test of the SIS to show conformance with the Safety Requirement Specifications. The PSAT shall include, but may not be limited to, confirmation of the following:*
>
> *(a)   SIS communicates (where required) with the Basic Process Control System or any other system or network.*
>
> *(b)   Sensors, logic, computations, and final control elements perform in accordance with Safety Requirement Specifications.*
>
> *(c)   Safety devices are tripped at the setpoints as defined in the Safety Requirement Specifications.*
>
> *(d)   The proper shutdown sequence is activated.*
>
> *(e)   The SIS provides the proper annunciation and proper operation display.*
>
> *(f)   The accuracy of any computations that are included in the SIS.*
>
> *(g)   That the system total and partial reset functions as planned.*
>
> *(h)   Bypass and bypass reset functions operate correctly.*
>
> *(i)   Manual shutdown systems operate correctly.*
>
> *(j)   Test interval is documented in maintenance procedures consistent with SIL requirements.*
>
> *(k)   SIS documentation is consistent with actual installation and operating procedures.*
>
> *A PSAT shall be satisfactorily completed prior to the introduction of hazards the SIS is designed to prevent or mitigate.*
>
> *Accuracy of calibration of test instruments used in the PSAT shall be consistent with the application. For example, the margin between the SIS setpoint and the hazardous process condition may be used to determine the required accuracy.*

Documentation to substantiate completion of the Commissioning and PSAT shall be completed prior to the introduction of hazards the SIS is designed to prevent or mitigate.

As a minimum, this documentation shall include the following:

(a)   Identification of the SIS that has been tested.

(b)   Confirmation that Commissioning is complete.

(c)   Date the PSAT was performed.

(d)   Reference to the procedures used in the PSAT.

(e)   Authorized signature that indicates PSAT has been satisfactorily completed.

11.6.1   Documentation Required for the PSAT

The documentation required to support the PSAT depends on the complexity of the safety system and which documents were originally prepared by the design team. Because the complete system is being tested, no single drawing will suffice. It is essential that a detailed procedure be prepared and followed. The following documentation is usually required to support the PSAT:

1. PSAT checkout procedures
2. copy of safety requirement specifications (SRS)
3. PES program listing/printout
4. a block diagram of the overall system
5. a list of inputs and outputs, complete with physical addresses
6. P&IDs
7. instrument index
8. specification sheets for all major equipment including manufacturer, model, and options
9. loop diagrams
10. electrical schematics
11. DCS configurations for any SIS inputs or outputs
12. cause-and-response diagrams and/or Boolean logic diagram
13. drawings to indicate the locations of all major equipment
14. junction box and cabinet connection diagrams

15. drawings to indicate the interconnections and termination of all wires

16. pneumatic system tubing diagrams

17. vendor standard equipment documentation, including (but not limited to) specifications, installation requirements, and operating manuals

11.7 Training

The process of training personnel for the safety systems is usually part of the overall training for the complete plant. It is essential that the criticality of the safety systems be well recognized and that a special commitment is made to ensure that personnel are well trained on the safety systems.

Because of the need for refresher training, the training for the safety system should start with the preparation of the safety requirement specifications and continue for the life of the plant. The key individuals to be trained are operations and support (including vendor) personnel.

The main training topics should be as follows:

- understanding the safety requirement specifications
- configuring and programming the PES and other electronic devices
- locating SIS equipment in the control room and field
- SIS documentation and location
- special maintenance procedures to be followed
- online testing procedures
- response to upset situations
- response to emergency shutdown situations

OSHA regulation CFR 1910.119 also defines requirements for training with respect to safety systems. The requirements of this document should be reviewed. The PES vendor support personnel are the individuals who will be called upon to provide final support. Find out who they are and how to contact them and ensure that they are knowledgeable about your system and have the capabilities to provide the required support.

The opportunities for providing effective training include the following:

- SRS preparation
- FAT

- logic simulation

- special classroom training

- site training manuals and systems

- PSAT

- start-up

- ongoing refresher training

11.8   Handover to Operations

The PSAT completion should be used as the basis for handing over the system to operations. Each interlock should be signed off to confirm that all tests have been completed successfully. If there are any outstanding issues, they must be reflected in the document with clear indication given regarding the following:

- What impact the deficiency has on the SIS operation.

- When it will be completed.

- Who has the responsibility for completion.

If operations feels that these outstanding deficiencies have the potential to create a hazard, then the start-up should be deferred or rescoped until they are corrected. Operations must be fully satisfied that they have been fully trained to operate the system.

11.9 Start-Up

During the initial start-up of a new plant the safety systems are likely to operate more often than they will for the rest of the plant's life. Start-up is probably the most hazardous phase in the life of the plant. The main reasons for this are as follows:

1. Even if the new plant is 100 percent identical to an existing one, there will always be certain aspects that are unique. Only after extended operation and experience will the actual operation be fully understood.

2. Changes are usually requested by operations to bypass or modify the trip setting for interlocks during the start-up period. Although this is expected, it is essential that a change procedure be implemented and rigorously enforced. The need for and impact of the change have to be analyzed thoroughly. (Refer to Chapter 13, "Managing Changes to a System.") Bypasses have been known to be in the enabled state for months after start-up. One common method for creating bypasses to avoid start-up trips is to force out-

puts from the PES. A hardwired alarm with ringback should be installed to monitor the forced outputs.

3. Operations may not be fully experienced at handling upset situations, thus escalating to a shutdown (S/D).

4. Some systemic problems (design and installation) may now be manifesting themselves and leading to trips.

5. Start-up is still within the "infant mortality" period for the PES and field devices. The component failure rates for the devices are highest at this time.

6. There is considerable activity during the start-up period, hence the potential for human error is very high.

11.10   Post-start-up Activities

The main activities remaining after start-up are documentation, final training, and the resolution of any deficiencies. A schedule should be prepared so that the completion of these activities is not prolonged.

A more comprehensive list of the remaining post-start-up activities includes the following:

- preparation of as-built drawings
- preparation of special maintenance documentation and procedures
- issuing the list and resolving the deficiencies
- training remaining operations and support personnel
- establishing online testing programs
- performing other preventive maintenance programs
- reviewing diagnostics within the system periodically, especially in the PES

Summary

In this chapter we have reviewed all activities from the completion of the design of the safety system to its successful operation. The key activities are as follows:

- factory acceptance test (FAT) of the logic system
- PES logic simulation
- installation of the complete system

- installation checks

- pre-start-up acceptance test (PSAT)

- training of operations and support personnel

- handover to operations

- start-up

- post-start-up activities

The overall objective of these activities is to ensure that the safety systems are installed in accordance with the detail design and perform in accordance with the safety requirement specifications and that personnel are adequately trained to operate and support the systems.

References

1. American Institute of Chemical Engineers, *Guidelines for Safe Automation of Chemical Processes* (New York: AIChE, Center for Chemical Process Safety, 1993), ISBN 0-8169-0554-1.

2. ANSI/ISA, *Application of Safety Instrumented Systems for the Process Industries*, ANSI/ISA S84.01 (Research Triangle Park, NC: ISA, 1996), ISBN 1-55617-590-6.

12

FUNCTIONAL TESTING

12.1 Introduction

Functional testing is an essential activity that has to be carried out periodically to verify the integrity of safety instrumented systems (SIS) and to ensure that the target safety integrity level (SIL) is met. The test must include the complete system, that is, sensors, the logic solver, the final element, and associated alarms, and must be based on clear and well-defined objectives, responsibilities, and written procedures.

The tests must be regarded as a normal preventive maintenance activity. Without periodic functional testing a SIS cannot be expected to function satisfactorily. SISs will fail, hence the systems *must* be tested in order to find inhibiting faults that would prevent the system from responding to a true demand. Sensors and electronic components can fail energized, and valves can stick. Do *not* assume that all modern electronic systems have thorough, automatic diagnostics. While some do, many do not. Just because all the green LEDs are on does *not* mean the system is fully functional!

The testing can be automatic or manual and can include hardware and/or software testing. Testing software is a controversial issue. One should realize that mere testing cannot detect all software errors. As we stated earlier, most software errors can be traced back to the requirements specification. Therefore, testing software against the requirements specification will not reveal all errors.

This chapter focuses on the periodic manual online testing of the mechanical devices.

It is not unusual to encounter safety systems in plants that have never been tested after the initial installation. When these installations are eventually tested we usually find an extraordinarily high rate of unsafe dormant failures in the safety system-- most of them associated with field devices and each of them capable of contributing to a potential accident. The main reasons this practice still persists are as follows:

1. SISs are treated as basic process control systems (BPCSs), hence the assumption that all SIS faults will be revealed. In Chapter 3, "Process Control versus Safety Control," the differences between BPCSs and SISs were emphasized. One of the key differences is the manner in which SISs fail. Unlike BPCSs, SISs can fail in two separate but distinct modes, that is, (1) fail-safe causing a nuisance trip or (2) fail-to-danger, in which case a trip will not occur if a demand is placed on the system. BPCSs do not normally exhibit fail-to-danger failure modes since most failures are easily identified. The primary objective of online functional testing is to identify fail-to-danger faults.

2. During the design phase, well-defined criteria for testing were not established and even if they were established the expected test frequencies are not being communicated to the maintenance or operations group for implementation.

3. Adequate procedures and facilities are not provided for carrying out the functional testing, so although it is recognized that the need exists functional testing is not done. In this case, the testing is usually deferred until a major maintenance shutdown of the plant or unit occurs. As a result, the test frequencies may be inadequate.

These items would be addressed if the life cycle model outlined in Chapter 1 were followed. The model described there addresses the need for and procedures involved in functional testing during the specification, design, operations, and maintenance phases.

The online functional testing of safety instrumented systems is not a trivial task. A tremendous amount of training, coordination, and planning is required to implement such tests and to prevent an actual process trip from occurring during the tests. Trips can occur, creating a hazardous situation and production losses. Because of this, one may encounter some resistance by operations personnel to getting involved with functional testing. Statements like "Not on my shift" or "If it ain't broke, don't fix it" will sometimes be heard.

One secondary benefit of online testing is that, once started, it increases the process and maintenance personnel's understanding of and confidence in the SIS.

12.2 Why Test?

The need for testing SIS is reinforced by considering the following:

- OSHA requirements
- Improvement of SIS performance
- S84.01 requirements
- S91.01 requirements

12.2.1 OSHA Requirements

OSHA's Process Safety Management (PSM) legislation (29 CFR 1910.119) requires employers to "establish maintenance systems for critical process related equipment including written procedures, employee training, appropriate inspections, and testing of such equipment to ensure ongoing mechanical integrity." The OSHA legislation further stipulates that "inspection and testing procedures shall follow recognized and generally accepted good engineering practices."[1]

12.2.2 Testing Improves SIS Performance

To better understand the necessity and rationale for functional testing, let us consider a very simple high-pressure SIS system consisting of a pressure transmitter, relay logic, and single trip valve (see Figure 12-1).

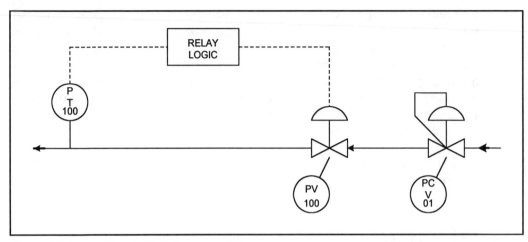

Figure 12-1   Simple High Pressure SIS

For the system shown in Figure 12-1, the probability of failure on demand (PFD_{avg}) is the unavailability of the SIS and the average probability that the SIS will not respond to a demand. For an interlock system with no redundancy or diagnostics (1oo1), this is expressed as follows:

$$PFD_{avg} = 0.5 * \lambda_d * TI$$

where:

λ_d =   Fail-to-danger rate of the entire system

λ_d = $\lambda_d(sensors) + \lambda_d(logic\ solver) + \lambda_d(final\ devices)$

TI   =   proof-test frequency

From this equation it is obvious that the proof-test frequency (TI) is as important a factor in establishing the PFD_{avg}/unavailability of the SIS in Figure 12-1 as is the fail-to-danger rate of the hardware components. Theoretically, if a system is never tested, TI$\Rightarrow \infty$, hence $PFD_{avg} \Rightarrow 1$. Therefore, the system has a very high probability of being unavailable when a demand occurs.

12.2.3 ANSI/ISA S84.01-1996 Requirements for Functional Testing

The following requirements have been established in ANSI/ISA S84.01-1996 for functional testing:[2]

Mandatory Sections

Clause 4. Safety Life Cycle

4.2.5 Performance is improved by the addition of redundancy, more frequent testing, use of diagnostic fault detection, and use of diverse sensors and final control elements, etc.

Clause 5. Safety Integrity Requirements

Safety integrity requirements shall include the following.

5.4.3 Requirements for maintenance and testing to achieve the required SIL.

Clause 7. SIS Detailed Design

7.9 Maintenance or testing design requirements.

7.9.1 The design shall allow for the testing of the overall system. It shall be possible to test final element actuation in response to sensor operation.

7.9.2 When online functional testing is required, test facilities shall be an integral part of the SIS design to test for covert failures.

7.9.3 When test and/or bypass facilities are included in the SIS, they shall conform with the following:

a) SIS shall be designed in accordance with the maintenance and testing requirements defined in the Safety Requirement Specifications.

b) The operator shall be alerted to the bypass of any portion of the SIS via an alarm and/or operating procedure.

c) Bypassing of any portion of the SIS shall not result in the loss of detection and/or annunciation of the condition(s) being monitored.

Clause 9.   SIS Operation and Maintenance

9.7   Functional testing.

Not all system faults are self-revealing. Covert faults that may inhibit SIS action on demand can only be detected by testing the entire system.

9.7.1   Periodic Functional Tests shall be conducted using a documented procedure.

9.7.2   The entire SIS shall be tested including the sensor(s), the logic solver, and the final element(s) (e.g., shutdown valves, motors).

9.7.3   Frequency of functional testing.

9.7.3.1 The SIS shall be tested at specific intervals based on the frequency specified in the Safety Requirement Specifications.

9.7.3.2 At some periodic interval (determined by the user), the frequency(s) of testing for the SIS or portions of the SIS shall be re-evaluated based on historical data, plant experience, hardware degradation, software reliability, etc.

9.7.3.3 Any change to the application logic requires full functional testing. Exceptions to this are allowed if appropriate review and partial testing of changes are done to ensure the SIL has not been compromised.

9.7.4   Functional testing procedures.

9.7.4.1 A documented functional test procedure, describing each step to be performed, shall be provided for each SIS.

9.7.5   On-line functional testing.

9.7.5.1 Procedures shall be written to allow on-line functional testing.

9.8   Documentation of functional testing.

9.8.1   A description of all tests performed shall be documented. The user shall maintain records to certify that tests and inspections have been performed.

Nonmandatory Sections

Annex B

B.15   *Functional test interval.*

B.15.1 *The frequency of functional tests should be consistent with applicable manufacturer's recommendations and good engineering practices, and more frequently if determined to be necessary by prior operating experience.*

B.15.2 *The functional test interval should be selected to achieve the Safety Integrity Level (SIL).*

B.15.3 *ISA-dTR84.02 illustrates various methods to determine the functional test interval.*

12.2.4   ANSI/ISA S91.01-1995 Requirements for Functional Testing

The following requirements have been established in ANSI/ISA S91.01-1995 for functional testing:[3]

4.2   *Maintenance and testing*

4.2.1   *All emergency shutdown systems and safety critical controls shall be periodically tested and maintained in accordance with user system test procedures taking into account system manufacturer recommendations.*

4.2.2   *The periodic tests of the emergency shutdown systems and safety critical controls shall contain the following minimum documentation:*

a) *Date of inspection;*

b) *Name of person who performed the test;*

c) *Serial number or other unique identifier of the equipment;*

d) *Results of the test as compared to user-defined acceptance criteria; and*

e) *A description of the test(s) performed.*

12.3 General Guidelines for Functional Testing

The following general guidelines should be considered for online functional testing:

- Online test facilities shall be provided for the testing of all safety instrumented systems (SIS) *if the equipment cannot be taken out of service to satisfy the required testing frequency.* The testing is required to confirm the correct operation of the system. All devices should be accessible for online testing.

- Procedures for online testing must be developed and followed prior to and during the commissioning of the SIS.

- The frequency of online testing must satisfy the SIS safety integrity requirements. The SIS designer must clearly establish the frequency of testing for all devices as part of the system design.

- Where it is not possible to test all components, for example, valves and motor starters, in the SIS on line, those devices must be designed with adequate availability and redundancy so that they can be tested during the turnaround of the plant. The design test frequency for the devices must be at least equal to the expected turnaround time for the plant.

- The testing facilities must enable the functional tests to be carried out easily with no interruption to the normal process operation.

- Documented test procedures with clearly defined responsibilities shall be used for all testing, and the test results shall be formally documented.

- Testing shall include sensors, logic, final devices, power supplies, and associated alarms.

- Field sensors are to be provided with local indicators to monitor process variables in the field and to view the simulated input signals during the test.

- All systems must be tested just prior to the start-up of any new facilities to ensure that inadvertent changes made during the commissioning are identified.

- The use of bypass switches for functional testing should be minimized. Whenever such switches are provided, the input signals for data logging and alarming should not be bypassed. A continuous hardwired flashing alarm light should be provided on a main annunciator to indicate that a bypass switch has been activated.

- The forcing of inputs or outputs in the logic system to simulate inputs or to activate outputs must be avoided. PES forcing should be restricted to maintenance testing when the complete safety system is out of service.

- Special procedures and tagging must be in place and followed before any bypasses or jumpers are installed.

- The effectiveness of the testing program must be reviewed annually. During the review the failures, testing frequencies, procedures, and problems encountered must be reviewed.

- The SIS must be retested after any repairs or modifications are carried out.

12.3.1   General Comments on Guidelines

It is not necessary to test all the elements of a safety system at the same time. Some devices can be tested on a more or less frequent basis reflecting their different failure rates. The PFD_{avg} equation is Section 12.2 can be rewritten as follows:

$$PFD_{avg} = 0.5(\lambda_{d(Sensors)}TI_S + \lambda_{d(Logic\ solver)}TI_L + \lambda_{d(Final\ devices)}TI_V)$$

where:

TI_S	=	Proof-test frequency for sensors
TI_L	=	Proof-test frequency for logic solver
TI_V	=	Proof-test frequency for final devices

Since the final devices usually contribute to the highest failures of safety systems, they can be tested at a more frequent rate to lower the PFD_{avg}. Don't therefore test every device because one element in the system requires testing. There must be a justification and rationale for every device tested. For example, if a temperature transmitter signal is connected to a safety system and the dangerous failure rate for the transmitter is very low, it may not be justifiable to set up test facilities and procedures to test the transmitter at the same frequency as the final elements.

During the test, the safety system may be off line or its trip capabilities severely reduced. This impacts the PFD_{avg} for the system. We should therefore attempt to optimize the test frequency. In a 1985 study, David Lihou and Z. Kabir discuss a statistical technique for optimizing the test frequencies of field devices based on past performance.[4]

During the testing and pre-start-up phase of any project, modifications are usually made to the safety system to get it "operational." These modifications can be changes to the set point or to the timer settings, the installation of jumpers or bypasses, or the disabling of trips. These "modifications" have to be controlled and as part of the plant start-up the system should be retested to verify that its operation is consistent with the safety requirement specifications (SRSs).

The testing of trip valves usually verifies whether the valves would open or close within a certain time. It is difficult to check whether the valve has internal mechanical problems. If this function is critical, additional facilities, for example, redundant trip valves, may be required.

12.4 Establishing Testing Frequency Requirements

There are no hard-and-fast rules for determining manual test intervals. One cannot simply say, for example, that SIL 3 systems must be tested monthly, SIL 2 systems quarterly, and SIL 1 systems yearly. Some companies have established standards that stipulate that SISs be tested at least once per year by trained personnel. These criteria may be plausible for establishing certain minimum requirements, but they may lead to an oversimplification of the standards in that there may be systems that require more frequent testing. For each system, the frequency of tests is a variable that will depend upon the technology, configuration, and target level of risk.

We saw the impact of manual test intervals on overall system performance in the modeling calculations of Chapter 8. It is possible to solve the formulas in reverse to determine the required test intervals in a quantitative manner. If spreadsheets or other programs are able to plot system performance based on the manual test interval as a variable, then the resulting plots will show the required manual test intervals based upon the desired level of system performance. Many people put in redundant systems specifically so they do *not* have to test them as often.

When determining the required manual test interval in a quantitative way, it is important to recognize the impact of the manual test duration. For example, when a simplex system is tested, it is normally necessary to place the system in bypass (so it does not shut down the process during testing). What is the availability of the system when it is in bypass? Obviously, the availability is zero. This should be factored into the calculations. During the test, a simplex system is reduced to zero, a dual system is reduced to simplex, and a triplicated system is reduced to dual. Formulas were given in Chapter 8 to account for this. This means there is an optimum manual test interval for any system.

It must be recognized that the test intervals that are established may be based on theoretical failure rates that may not reflect actual operating conditions. The test intervals and failures must therefore be reviewed periodically to ensure that the actual failure rates are used to establish the test frequencies.

In cases where failure data cannot be obtained to establish test frequencies, the test intervals recommended by ANSI/ISA S84.01-1996 should be used, unless other data is available to support testing less frequently.

12.5 Responsibilities for Testing

It is essential that those involved in the functional testing of SIS be very knowledgeable about both the process and the SIS design. In addition to verifying the integrity of the SIS, the functional testing provides valuable training for those performing the test and orientation training for new employees witnessing the test.

The testing can be performed by the following people:

- process operating personnel
- maintenance personnel
- jointly by operations and maintenance personnel

In most plants, process personnel are the "owners" of the plant and are ultimately responsible for its safe operation. For this reason, the onus for testing should be placed on them. The testing facilities and procedures should be such that the testing can be completed satisfactorily by the operations personnel. Maintenance personnel should only be involved with functional testing if any special action is required by them, for example, installing jumpers or a temporary bypass. In online functional testing, every attempt should be made to maximize the use of process operating personnel and to minimize the use of maintenance personnel.

12.6 Test Facilities and Procedures

The procedure for testing will vary considerably with the type of field device, the testing requirements, and facilities provided for the testing. All components that are part of the SIS should be tested, and all devices should be tested while they are in service with an absolute minimum of changes to the system. For example, to provide adequate online testing facilities for the high-pressure safety system shown in Figure 12-1 the following steps must be taken:

1. For PT-100, valves are required to isolate the process connection and to pressurize the transmitter.

2. For PV-100, blocks and bypasses have to be installed for the trip valve.

Figure 12-2 shows a typical installation.

To test the system shown in Figure 12-2, the bypass valve BV-100 would be opened and the signal to the transmitter changed to simulate a trip. The main trip valve PV-100 should close at the trip setting. Also, some form of feedback (e.g., limit switches) could be used to verify that the trip valve has in fact closed fully.

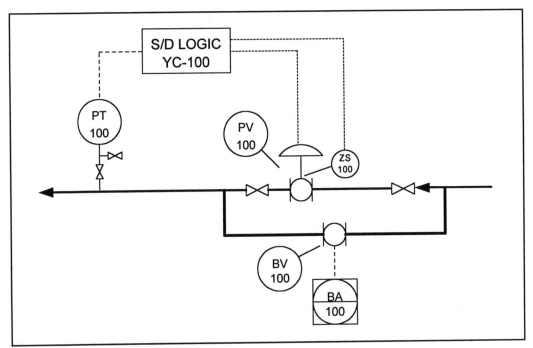

Figure 12-2   Typical On-line Testing Facilities

Some general guidelines for testing are as follows:

- Low-pressure sensing device:Close the main process isolation valve and vent signal to the transmitter or switch until the device activates.

- High-pressure sensing device:Close the main process isolation valve and inject a simulated signal to the device until the device activates.

A common practice is to use local indicating pneumatic on/off controllers as the input device. The output signal from the on/off controllers is either 3 or 15 psi. A pressure switch set at 9 psi would act as the input to the logic system. By changing the set point of the on/off controllers the input devices can easily be tested.

If a valve opens for a shutdown, a normally chained open block valve could be installed upstream of the actuated valve. This block valve could then be closed, and the shutdown valve could be opened for testing. After testing, everything must be returned to the original positions.

The use of manual bypass switches to disable the trip valve from closing should be avoided and only used if it is not practical to install a bypass around the trip valve. In some cases, it is impossible to test or trip the final device, for example, a compressor, a pump, or a large valve. In these cases, some form of bypass has to be installed to ensure that the final device won't trip when the primary device and logic is tested. In the case of a compressor, the final shutdown relay can be energized and the signal from the shutdown relay to the compressor starter bypassed.

If manually operated bypass switches have to be installed, only one switch should be operated at a time. Trip overrides are usually provided to permit the plant to start up. Alarms associated with these overrides must be designed in much the same way as online testing bypass switches. It is preferable for the overrides to be controlled by timers so the trip system's protective features are reestablished after a fixed time.

Redundant systems also have to be tested to ensure that the fault-tolerant features are fully functional. Any defect identified during the test must be immediately reported and repairs scheduled based on the criticality of the system.

12.7 Documentation

All operating, test, and maintenance procedures must be fully documented and made available to all personnel. (Simply having procedures, however, does not ensure that they will be *read*.) It is a good idea to involve operating personnel in the development of the test procedures, for no one knows more than they how things actually need to be and can be done.

All testing should leave a paperwork trail, one that can be audited by an independent person or group at a later time. Each shutdown loop requires a separate procedure.

12.7.1 Trip Test Procedure Documentation

All procedures for online testing must contain and list the information required by the individuals carrying out the testing. Also, a step-by-step list of the actions required to carry out the test should be clearly documented.

A suggested list of the key items to be included in a test procedure is as follows:

TRIP TEST PROCEDURE   **PROCEDURE #:**_____

1.0   <u>PURPOSE</u>

Define clearly the purpose and objective of the test.

2.0   <u>RESPONSIBILITY FOR TEST</u>

One individual must have overall responsibility for the test. Personnel required to assist the responsible individual must also be listed.

3.0   <u>TOOLS AND OTHER ITEMS REQUIRED</u>

Any special tools and other facilities required must be listed. We must also ensure that the tools are in good operating condition.

4.0   <u>TRIP CHECK FREQUENCY</u>

State the frequency with which the test is to be completed.

5.0   HAZARDS

Any special hazards that are part of the test or that can be created as a result of the test must be well documented.

6.0   REFERENCE INFORMATION

A simplified sketch, and description of the trip system operation should be part of this procedure. Also, relevant loop drawings, spec sheets, schematics, reference or vendor manuals, or material data sheets should be listed here.

7.0   DETAILED STEP-BY-STEP TEST PROCEDURE

The detailed step-by-step test procedure must ensure that the following is verified:

- *correct functionality of input, logic, output, reset, and alarm devices*
- *the speed of response of system is acceptable*
- *"as found" versus expected values should be documented*

RESULTS/COMMENTS/DEFICIENCIES

Document results of the test. Any comments or deficiencies identified during the test must be listed.

TRIP CHECK COMPLETED BY _____   **DATE:**_____

Summary

Safety instrumented systems (SISs) have to be tested to verify their integrity and to ensure that the target safety integrity level (SIL) is met. Testing is also a requirement of OSHA regulations and the ANSI/ISA S84.01-1996 standard. The test requirements must include the complete system, that is, sensors, the logic solver, the final element, and associated alarms, and must be based on clear and well-defined objectives, responsibilities, and written procedures.

In this chapter we reviewed the following topics:

- why testing is required
- general guidelines for functional testing
- how to establish test frequencies
- responsibilities for testing
- test facilities and procedures, and
- documentation required for testing

References

1. OSHA, *Process Safety Management of Highly Hazardous Chemicals, United States Code of Federal Regulations*, 29 CFR 1910.119 (Washington, DC: Government Printing Office).

2. ANSI/ISA, *Application of Safety Instrumented Systems for the Process Industries*, ANSI/ISA S84.01-1996 (Research Triangle Park, NC: ISA, 1996), ISBN 1-55617-590-6.

3. ANSI/ISA, *Identification of Emergency Shutdown Systems That Are Critical to Maintaining Safety in Process Industries*, ANSI/ISA S91.01-1995 (Research Triangle Park, NC: ISA, 1995), ISBN 1-55617-570-1.

4. David Lihou and Z. Kabir, "Sequential Testing of Safety Systems," *The Chemical Engineer*, December 1985.

13

MANAGING CHANGES TO A SYSTEM

"The only constant in any organization is change."

13.1 Introduction

Making changes to processes, process controls, safety systems, equipment, procedures, and the like for industrial plants is inevitable. Changes are required for many reasons, for example, technology, quality, improvements, or equipment malfunction.

OSHA regulations (OSHA 29 CFR 1910.119) require employers to "establish and implement written procedures to manage changes (except 'replacements in kind') to process chemicals, technology, equipment, and procedures; and changes to facilities that affect a covered process."[1]

Similarly, ANSI/ISA standard S84.01-1996 mandates that

- "If modifications are proposed, their implementation shall follow a Management of Change (MOC) procedure. The appropriate steps in the Safety Life Cycle shall be repeated to address the safety impact of the change."

- For any changes made to an operating safety instrumented system (SIS) "a written procedure shall be in place to initiate, document, review the change, and approve changes to the SIS other than 'replacements in kind.'"[2]

This chapter focuses on the management of change (MOC) as it relates to safety systems, with specific reference to the OSHA and ANSI/ISA S84.01-1996 requirements. The MOC procedure will ensure that all changes are made in a safe, consistent, and well-documented manner.

13.2 The Need to Manage Changes

Changes made to a process, no matter how minor, may have severe safety implications. Take one major process accident—the Flixborough explosion in England—as an example. In June 1974, there was a massive explosion at the Nypro factory in Flixborough that fundamentally changed the way in which safety is managed by all industries.

The Nypro plant produces an intermediate used to make nylon. The unit that exploded oxidized cyclohexane with air to produce cyclohexanone and cyclohexanol. The process consisted of a series of six reactors, each holding twenty tonnes and each partly connected with a flexible joint. Liquid overflowed from one reactor to the next and fresh air was added to each reactor. A crack and leak was discovered on one of the reactors, and it was removed for repair. To maintain production a temporary pipe bypass link was installed in its place.

A qualified engineer was not consulted. The only layout made of the change was a full-scale chalk drawing found on the workshop floor. The system was pressure tested once it was installed, but it was pneumatic rather than hydraulic and conducted to 127 psig rather than the relief valve pressure of 156 psig. The design violated British standards and guidelines.

The temporary pipe was okay for two months until there was a slight rise in pressure. It was less than the relief valve pressure, but it caused the temporary pipe to twist. The bending movement was strong enough to tear the flexible connection and eject the temporary pipe, leaving two 28-inch pipe openings.

The resulting gas cloud found a source of ignition in the plant and ignited. The resulting explosion resulted in twenty-eight deaths and thirty-six injuries and was estimated to be the equivalent to fifteen to forty-five tons of TNT. The plant was devastated, buildings within a third of a mile were destroyed, windows were shattered up to eight miles away, and the explosion was heard thirty miles away. Flames from the resulting fire continued to burn for ten days, hindering rescue work at the site.

The main lesson learned from Flixborough is that all modifications must be designed and reviewed by qualified personnel. The men installing the pipe did not have the expert knowledge required.

Stringent management of change procedures should be followed to analyze the impact of any and all changes. What may seem to be insignificant changes to one person may in fact represent considerable risk. This process requires input from people who understand the instrumentation, the control system, the process, and the mechanical/structural systems. It is important to leave an auditable trail.

To make any system changes, it is important that *original* safety-related design bases be used. Engineers of both the hardware and software need to specify the constraints, assumptions, and design features so that maintainers do not accidentally violate the assumptions or eliminate the safety features. Decision makers need to know why safety-related decisions were made originally so they do not inadvertently undo them.[2]

Changes can inadvertently eliminate important safety features or diminish the effectiveness of controls. Many accidents have been attributed to the fact that the system did not operate as intended because changes were not fully analyzed to determine their effect on the system.[2]

13.3 When Is MOC Required?

The MOC would apply to all elements of the safety system, including the field devices, logic, final elements, alarms, users' interface, and application logic. ANSI/ISA S84.01-1996 states that an MOC procedure could be required as a result of the following modifications[3]:

1. to the operating procedure

2. those necessary because of new or amended safety legislation

3. to the process

4. to the safety requirements specification (SRS)

5. to fix software or firmware errors

6. to correct systematic failures

7. as a result of a failure rate higher than desired

8. resulting from increased demand rate on the SIS

9. to software (embedded, utility, application)

Additional guidelines and clarifications requiring an MOC procedure are as follows:

- The change will show a deviation from the associated P&ID.

- A component of the SIS is being removed.

- The online testing or maintenance procedure for the SIS is being modified.

- The documentation format for the SIS is being changed.

13.4 When Is MOC Not Required?

An MOC procedure is not normally required for the following situations:

- Changes deemed as "replacements in kind."

  With regard to safety system components, "replacements in kind" does not have the same meaning as replacing, say, an electric motor. In replacing the electric motor we can install another manufacturer's motor if the horsepower, voltage, frequency, speed, and frame size are the same. To replace an SIS component--for example, a transmitter--the failure rate and failure modes of the new transmitter are also critical since they directly affect the performance of the SIS. For SIS, "replacements in kind" should be "like for like" or a preapproved replacement that satisfies the original specification.

- Any changes that are within the safety requirements specification.

- Repair of equipment to original state after failure.

- Calibration, blowdown, or zeroing of field instruments that are part of the SIS.

- Range changes, alarm, or trip changes within design or operation conditions as outlined in the safety requirements specifications.

- Any changes covered by another internal corporate procedure that specifically states that the MOC need not be implemented for the specific change.

13.5 Considerations

13.5.1   OSHA and ANSI/ISA S84.01-1996 Considerations

- OSHA and/or ANSI/ISA S84.01-1996 state that the following considerations are to be addressed prior to any change:

  1. the technical basis for the proposed change

  2. the impact of the change on safety and health

  3. modifications for operating procedures

  4. necessary time period for the change

  5. authorization requirements for the proposed change

  6. availability of memory space

  7. effect on response time

  8. online versus offline change and the risks involved

- The review of the change shall ensure that

  1. the required safety integrity has been maintained; and

  2. personnel from appropriate disciplines have been included in the review process.

- Personnel affected by the change shall be informed of the change and trained prior to the implementation of the change or the start-up of the process, as appropriate.

- All changes to the SIS shall initiate a return to the appropriate phase (the first phase affected by the modification) of the safety life cycle. All subsequent safety life cycle phases shall then be carried out, including the appropriate verification that the change has been carried out correctly and has been documented. The implementation of all changes (including application software) shall adhere to the previously established SIS design procedures.

13.5.2   Other Considerations

- Software changes should be treated as seriously as changes to the plant or process and should be subjected to similar control. Any changes to the software should require retesting of the complete logic associated with the changes.

- When a shutdown system is deactivated, management-of-change (MOC) procedures are required to assure that no other systems or processes will be affected. All decommissioning activities therefore require an MOC procedure.

13.6 MOC Procedure

The management-of-change (MOC) procedure should include the following:

Items to be Included in MOC Procedure	Comments
1. How to initiate or request change	Change approval forms need to be developed and included with the MOC procedure. The technical basis for the change should be included in the form. This form is usually generic for all site changes. Both temporary and permanent changes have to be included.
2. Approval of change request	The authorization/approval process for the change has to be established.
3. Responsibilities	Identify the individual(s) responsible for the following: - approving the change request - identifying the need to implement a MOC procedure, and - for ensuring that the procedure will be followed.
4. Hazard assessment	The individuals involved in the hazard assessment are usually not the same as the original HAZOP team. These individuals should be identified. More operational and maintenance personnel are usually involved in the assessment. The techniques required for the assessment must be identified based on the nature, size, and impact of the change.
5. Documentation	What documentation has to be updated? How will the documentation be updated?
6. Review and sign-off	Who will review the updated documentation and sign off?
7. Training	Everyone affected by the change has to be appraised and trained.
8. Construction	When and how will the changes be made in the field?
9. Checkout	A separate crew should be responsible for the checkout based on well-documented procedures.
10. Start-up	Prior to start-up the changes must be approved by operations.
11. Final completion	Upon successful start-up a final report should be signed off and issued.
12 Auditing and verification	An auditing and verification system has to be in place to verify that the procedure is being followed.

13.7 MOC Documentation

The following relevant documentation should be updated:

- application logic
- design documentation
- commissioning pre-start-up acceptance test procedure(s)
- SIS operating procedure(s)
- SIS maintenance procedure(s)
- functional test procedure(s)
- safety requirements specifications

Maintaining up-to-date and accurate documentation for safety systems is critical. Inaccurate documentation by itself creates a potential hazard.

ANSI/ISA S84.01-1996 also stipulates the following documentation requirements:

- "All changes to operating procedures, process safety information, and SIS documentation (including software) shall be noted prior to startup and updated accordingly."
- "The documentation shall be appropriately protected against unauthorized modification, destruction, or loss."
- "All SIS documents shall be revised, amended, reviewed, approved and be under the control of an appropriate document control procedure."

Summary

Making changes to processes, process controls, safety systems, equipment, procedures, and the like for industrial plants is inevitable. Changes are required for many reasons, for example, technology, quality, improvements, or equipment malfunction.

A management-of-change (MOC) procedure as outlined in ANSI/ISA S84.01-1996 and the OSHA 29 CFR 1910.119 regulations must be followed for all changes outside of the safety requirements specifications (SRSs).

The procedure must ensure that the changes are

- *specified* and *designed* by competent individuals
- *reviewed* by an appropriate team capable of addressing the technical basis for the change and identifying the impact of the change

- *inspected* and *fully tested* prior to putting the system back into operation

- fully *documented* in accordance with the site's documentation requirements

- *communicated* to affected employees so that they are adequately informed and trained.

References

1. *Process Safety Management of Highly Hazardous Chemicals, United States Code of Federal Regulations*, 29 CFR 1910.119 (Washington, DC: Government Printing Office, 1992).

2. Nancy G. Leveson, *Safeware - System Safety and Computers* (Addison-Wesley, 1995), ISBN 0-201-11972-2.

3. ANSI/ISA, *Application of Safety Instrumented Systems for the Process Industries*, ANSI/ISA S84.01-1996 (Research Triangle Park, NC: ISA 1996), ISBN 1-55617-590-6.

14

JUSTIFICATION FOR A SAFETY SYSTEM

14.1 Introduction

The life cycle model described in Chapter 1 emphasizes that the goal of process design is to have inherently safe plants. Safety instrumented systems (SISs) should only be installed if other noninstrumented layers cannot adequately mitigate the risks. The emphasis is always toward *not* having to provide safety instrumented systems to reduce the risks or prevent hazards.

If it is determined that a safety instrumented system is required then the integrity of the system has to be identified so as to adequately mitigate the risks; that is, for *each safety function* we have to identify whether a SIL 1, 2, or 3 is required.

With this life cycle model approach, the safety instrumented system is being installed primarily for safety reasons, and the assumption is that shutting the process down will address the safety concerns. The justification is very straightforward using this approach, provided that an acceptable procedure is in place to determine the SIL level. Techniques for determining SIL levels were discussed in Chapter 6. There aren't any justification issues with the life cycle model approach—the safety requirements have to be satisfied for legal, moral, financial, and ethical reasons.

Justifying a safety system purely from the point of view of safety is usually not an issue.

What is not adequately addressed in the life cycle model is the link between the safety, reliability, and the life cycle cost of the safety system. The safety issue addresses the integrity of the system in adequately responding to a demand. The reliability issue addresses the nuisance trip rate and the impact of nuisance trips on safety and cost. The life cycle cost analyzes the total cost of the safety system for its life.

A safety system has to be justified not only in terms of safety but also in terms of its reliability and life cycle cost.

Another issue facing control system designers is the use of the safety instrumented system, not in addition to but instead of other layers. Control system designers are being asked to analyze and recommend the possible installation of safety systems to replace other layers for cost, scheduling, or competitive reasons.

This chapter therefore looks at the following important issues and their impact on safety system justification:

- failure modes of safety systems
- responsibilities for justifications
- life cycle cost analysis
- optimizing safety, reliability, and life cycle costs

14.2 Safety System Failures

Throughout this book we have stressed that safety systems fail in two distinct modes: fail-safe or fail-to-danger. A discussion of the impact of these failures on safety system justification must be based on a consideration of the following four distinct failure modes:

1. *Dangerous undetected failure:* In this mode, there is no indication that the system has failed, and it will not respond to a demand.

2. *Dangerous detected failure:* In this mode, the system has failed, but the failure is annunciated to the operator through the internal self-

diagnostics of the system. In this scenario, the system will also fail to respond to a demand.

3. *Degraded mode due to a partial failure of a redundant system:* This failure is annunciated to the operator through the internal self-diagnostics of the system, but the safety shutdown functions are still active. In this scenario, the system will respond to a demand. This applies to redundant systems that are fault tolerant.

4. *Nuisance trip failure:* This is a failure that produces a nuisance or spurious trip. The plant has to be restarted once the problem is corrected.

14.2.1   Mode 1: Dangerous Undetected Failure

Mode 1 is the worst scenario. The probability of a dangerous undetected failure occurring is related to the PFD_{avg} and the safety integrity level (SIL) of the system.   The SIL determination should be based on the risk management or risk tolerance philosophy of the corporation or company. This policy must be well established and understood. If it is, the SIL determination is unlikely to be challenged by management. Whatever safety system architecture or redundancy is required to comply with the SIL determination is usually readily accepted. For this mode, the only way to determine that such a failure has occurred is by performing periodic manual online testing.

14.2.2   Mode 2: Dangerous Detected Failure

Mode 2 would obviously be the preferred mode if a dangerous failure were to occur. If the mean time to repair (MTTR) the system is, say, eight hours, and if a demand occurs every six months, then the probability of a hazard occurring within the eight-hour period is approximately one in five hundred.

Upon identifying such a failure, the process operator would follow established internal procedures for such failures. For some SIL 3 systems this may require an immediate manual shutdown of the process. Most likely, the procedures will require the operator to do the following:

1. Request that support personnel rectify the problem. If the problem is not rectified within a predefined time limit, the system will shut down automatically. This predefined time limit is usually set by approval agencies such as TÜV.

2. Monitor the trip parameters and the overall process closer and be prepared to take manual action to bring the process to a safe state upon abnormal conditions. This manual intervention further

reduces the probability that a hazard will occur due to the non-availability of the SIS.

The diagnostic capabilities of the safety system, including field devices, have to be part of the inherent design and must be well tested. Additional costs may be required to provide the diagnostics. This has to be justified.

14.2.3   Mode 3: Degraded Mode

Redundant systems have the capability to be fault tolerant and are widely used whenever both safety and availability/reliability issues need to be addressed. This is obviously the preferred failure mode since the safety shutdown capabilities of the system still exist while the problem is being rectified. Increased redundancy increases the initial installation costs, but the life cycle cost of the system may be less. The life cycle cost analysis in Section 14.6 shows the impact of redundancy on the total cost of a safety system.

14.2.4   Mode 4: Nuisance Trip Failure

The full impact of nuisance trips on processes isn't adequately addressed in the standards and hence is not analyzed and discussed at the required level. The following paragraph from OSHA's Process Safety Management (PSM) legislation (29 CFR 1910.119) summarizes the concerns about nuisance trips:[1]

> *In refining processes, there are occasionally instances when a piece of equipment exceeds what is deemed "acceptable," and interim measures are taken to bring the equipment back into conformance with safe operating parameters. Under (j)(4) it would be mandatory to immediately shut down the entire process upon discovery of such a situation. Shutdowns and startups are inherently dangerous operations which we try to avoid unless absolutely necessary. In addition, the life expectancy of certain components is directly affected by the number of cycles to which they are subjected. We feel that safety is promoted rather than diminished by keeping shutdowns to a minimum.*

In addition, safety shutdown systems, or parts of them, have reportedly been bypassed when there is an unacceptable level of nuisance trips and production is being severely affected. Accidents have been related to such bypasses. To reduce spurious trips, the $MTTF_{spurious}$ of the system has to be increased. This again impacts the cost and complexity of the system and has to be justified.

14.3 Responsibilities for Justification

Justifying a safety system purely from the point of view of safety, irrespective of cost or complexity, is usually not a problem. It is well understood that the safety requirements have to be satisfied. Issues usually arise, however, when it is recommended that the cost or complexity of the system be enhanced to improve its reliability so as to satisfy the nuisance trip requirements.

Justifying the enhancements to decrease the nuisance trips usually faces some opposition, for the following reasons:

- Such enhancements are usually perceived by many safety personnel as a production rather than a safety issue. This is a false perception.

- The objective of the safety system is to protect the plant. Unwarranted shutdown caused by the failures of the safety system is regarded as a control system issue that should be solved by control system personnel. Don't be surprised to hear a production manager say to a control systems engineer, "What do you mean, your shutdown system will shut down my plant at least once every year?"

- Project budgets are normally fixed at this stage, and there is a reluctance to spend extra dollars unless it is well justified.

- The data used to calculate the nuisance trip rate is often challenged.

Justifying a safety system that is above and beyond the safety requirements is usually the responsibility of the control systems engineer.

Although the benefits of a more reliable safety system may appear obvious, don't assume that everyone will be in agreement. Some common questions will be the following:

- "The last project didn't require such a complex safety system. Why are you now insisting on such complexity and extra cost?"

- "Can we upgrade the system later if we really have a problem?"

- "What other less costly options have you considered, and what is the best alternative?"

Obviously, there is a need to properly justify the expenditure for the safety system. Within any corporation, all projects are competing for fewer and fewer dollars. The impact of the expenditure on the bottom line will be closely scrutinized.

14.4 How to Justify

The most efficient way to justify the expenditure is to complete a life cycle analysis for the various options being considered. The life cycle costs reflect the total cost of owning the system. By calculating the life cycle costs, the various options can be analyzed in a more effective and consistent way. Table 14-1 describes the predominant costs incurred in the life of a safety system. The list is divided into (1) initial fixed costs, which are the costs to have the system designed, purchased, installed, commissioned, and operational, and (2) the annual costs, which are the maintenance and other ongoing costs associated with the system. To some extent, the costs reflect the items listed in the life cycle model (see Chapter 1).

14.5 Example

The following example will be used to develop the life cycle costs for two possible safety system solutions.

> Flammable materials A and B are fed continuously into a reactor vessel. The flow of materials A and B to the vessel are automatically controlled from the basic process control system (BPCS) in a fixed ratio. The set point of the primary flow controller A is set by the vessel level controller so as to maintain a fixed level in the vessel. The flow controller for feed A adjusts the set point of the flow controller for feed B to maintain the fixed ratio.

Figure 14-1 shows the basic process controls.

Cost Item	Comments
Initial Fixed Costs	
Safety classification	Once the HAZOP is completed and a safety system is required, the costs to complete the classification are the first costs associated with the system.
Safety requirements and design specifications	Reflect the manpower costs to complete the safety requirements specifications, conceptual design, and design specification.
Detailed design and engineering	This is the total cost for the complete detailed design and engineering package to be used for procurement, installation, testing, and start-up.
Sensors	Purchase cost of sensors.
Final elements	Purchase cost of valves and other final elements.
Logic system	Purchase cost of the logic system.
Miscellaneous: power, wiring, junction boxes, operators' interface	Costs for other miscellaneous equipment required to install and monitor the safety system.
Initial training	The cost of training for design, operations, and support personnel to design, install, and test the system.
FAT/installation/PSAT	Total costs for factory acceptance tests, equipment installation, and pre-start-up acceptance tests.
Start-up and correction	Most systems require some correction prior to full operation.
Annual Costs	
Ongoing training	Ongoing refresher training for operations and support personnel is required.
Engineering changes	These costs are usually quite significant because of the review requirements and documentation updates.
Service agreement	Programmable logic systems usually required a service agreement with the manufacturer to resolve "difficult" problems.
Fixed operation and maintenance costs	For example, utilities or preventive maintenance programs.
Spares	Critical spares as recommended by vendors.
Online testing	Testing carried out on a periodic basis by operations and support personnel.
Repair costs	Costs for repairing defective modules based on predicted failure rates.
Hazard costs	The hazard rate is a function of the unavailability of the system and the demand rate. This cost is based on the hazard impact.
Spurious trips costs	This cost is based on the $\text{MTBF}_{spurious}$ for the system and reflects the cost of lost production.
Present value for annual costs	This is the present value of the annual costs based on current interest rates and the predicted life of the system. These costs are added to the initial fixed costs to obtain the total life cycle costs. The PV is usually calculated by solving the following equation for each year: $PV_N = M/((1+R)^N)$, where: M is the future value, R is the interest rate, and N is the year. $PV = \Sigma PV_N$
Total Life Cycle Costs	Total safety system cost for the life of the system. This is the sum of the initial fixed costs and the present value of the annual fixed costs.

Table 14-1   Breakdown of SIS Costs

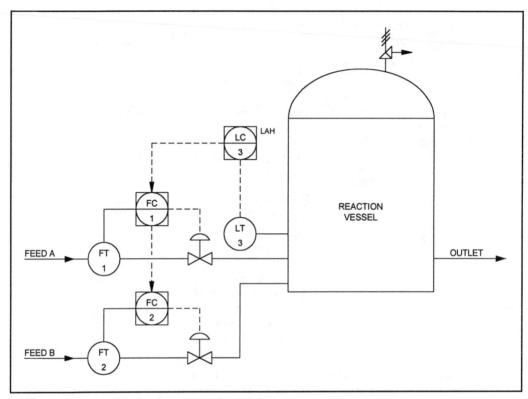

Figure 14-1   Basic Process Controls for Vessel

Continuing the example, as a result of a hazard review for the system the hazards shown in Table 14-2 were identified.

Hazard	Cause(s)	Consequence	Likelihood of Occurrence
Release of flammable gas into environment	Failure of BPCS	Fire, explosion $500K loss	Medium
Vessel failure	Failure of BPCS and relief valve	$750K loss	Low
Note: A nuisance trip costs $10K per trip.			

Table 14-2   Hazard Analysis

The following safety and protective instrumentation was recommended:

1. Install a high vessel pressure trip to close off feed A and B to the vessel.

2. Install a high vessel level trip to close off feed A and B to the vessel.

Based on this data, an SIL 2 is required. The following safety systems were proposed:

CASE 1

1. Sensors: Single transmitters
2. Logic: Relay logic
3. Valves: Single independent trip valves on each line

CASE 2

1. Sensors: Transmitters 2003 voting
2. Logic: Redundant programmable electronic system (PES)
3. Valves: Independent trip valves 1002 voting

14.5.1   PFD and Nuisance Trip Calculations for Case 1

The PFD and nuisance trip calculations for case 1 given in this section are illustrated in Figure 14-2.

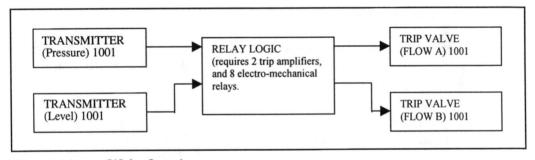

Figure 14-2   SIS for Case 1

The PFD and nuisance trip calculations for case 1 are as follows:

Failure data (failures/year)

Transmitter:	λ_d =	0.01	λ_s =	0.02
Valve and sol^n:	λ_d =	0.02	λ_s =	0.1
Trip amplifiers:	λ_d =	0.01	λ_s =	0.01
Mechanical relay:	λ_d =	0.002	λ_s =	0.02

Additional data

MTTR: 8 hours
Average demand rate:Every 6 months
Online test interval: Every 3 months

PFD calculations

The following assumptions have been made: (1) the MTTR is small enough to be insignificant, (2) λ_d includes all dangerous failures, hardware and systematic, (3) common cause problems are insignificant, (4) failure rates for auxiliaries, i.e., power supplies, cables, and the like, are negligible.

Note: For a dangerous failure, we only need to factor one transmitter, both valves, one trip amplifier, and two mechanical relays. Refer to Chapter 8 for the equations.

PFD (sensors)	=	$0.5 * 0.01 * 3/12$	= 0.00125
PFD (valves and sol^n)	=	$2 * 0.5 * 0.02 * 3/12$	= 0.0050
PFD (trip amplifier)	=	$0.5 * 0.01 * 3/12$	= 0.00125
PFD (mechanical relay)	=	$2 * 0.5 * 0.002 * 3/12$	= 0.0005
PFD (total)	=	0.008	

The maximum value required for SIL 2 is .01.

This system therefore satisfies the safety requirements.

Nuisance trip calculations

Note: To calculate the nuisance trip rate all components have to be considered, i.e., two transmitters, both valves with solenoids, two trip amplifiers, and eight mechanical relays.

$MTBF_{sp}$ (sensors)	=	$1/(2 * .01)$	= 50 yrs
$MTBF_{sp}$ (valves and sol^n)	=	$1/2 * .1$	= 5 yrs
$MTBF_{sp}$ (trip amplifier)	=	$1/(2 * .01)$	= 50 yrs
$MTBF_{sp}$ (mechanical relay)	=	$1/(8 * 0.02)$	= 6.25 yrs
$MTBF_{sp}$ (total)	=	2.5 years	

A nuisance trip is expected to occur, on average, every 2.5 years.

14.5.2 PFD and Nuisance Trip Calculations for Case 2

The PFD and nuisance trip calculations for case 2 given in this section are illustrated in Figure 14-3.

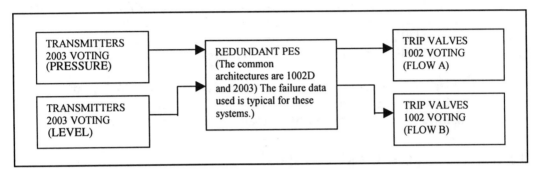

Figure 14-3   SIS for Case 2

Failure data (failures/year)

Transmitter: λ_d = 0.01 λ_s = 0.02
Valve and sol^n:λ_d = 0.02 λ_s = 0.1

Additional data

MTTR: 8 hours
Average demand rate:Every 6 months
Online test interval: Every 3 months

PFD calculations

The following assumptions have been made: (1) the MTTR is small enough to be insignificant, (2) λ_d includes all dangerous failures, hardware and systematic, (3) common cause problems are insignificant, (4) failure rates for auxiliaries, i.e., power supplies, cables, and the like, are negligible.

Note: For a dangerous failure, we only need to factor one 2003 transmitter system, both 1002 valve systems, and the PFD for the logic system.

PFD (sensors)   = $(.01)^2(3/12)^2$ = .000006
PFD (valves and sol^n) = $2 * .333 * (.02)^2(3/12)^2$ = .000016
Redundant PES: = .00001
PFD (Total)   = 0.000032

The maximum value required to SIL 2 is .01.

This system more than adequately satisfies the safety requirements.

Nuisance trip calculations

Note: To calculate the nuisance trip rate all components have to be considered, i.e., all transmitters, all valves with solenoids, and the safe failure rate for the PES.

$$
\begin{aligned}
MTBF_{sp} \text{ (sensors)} &= 1/(6 * .02 * .02 * (8/8760)) = {>}1000 \text{ yrs} \\
MTBF_{sp} \text{ (valves and sol}^n) &= 1/(2 * 2 * .1) = 2.5 \text{ yrs} \\
MTBF_{sp} \text{ (PES)} &= {>}1000 \text{ yrs} \\
MTBF_{sp} \text{ (total)} &= 2.5 \text{ yrs}
\end{aligned}
$$

A nuisance trip is expected to occur, on average, every 2.5 years. This is only because the valves are installed in a 1002 voting configuration. The transmitters and PES have little impact on the nuisance trips.

14.6 Life Cycle Cost Analysis

This analysis given in Tables 14-3 and 14-4 in this section is intended to demonstrate the technique for completing the life cycle cost analysis. The *data is very subjective* and is not intended to reflect the actual costs for any particular product or installation technique. Individuals will have to rely on data within their own organization for more accurate numbers.

	Material $	Labor $	Total cost $	Subtotal $
Initial Fixed Costs				
Safety classification		$2,000	$2,000	
SRS/design specifications		$5,000	$5,000	
Detailed design and engineering		$30,000	$30,000	
Sensors	$4,000		$4,000	
Final elements	$3,000		$3,000	
Logic system	$10,000		$10,000	
Miscellaneous: power, wiring, junction boxes	$5,000		$5,000	
Initial training		$5,000	$5,000	
FAT/installation/PSAT	$5,000	$25,000	$30,000	
Start-up and correction	$2,000	$8,000	$10,000	
Fixed costs subtotal				$104,000
Annual Costs				
Ongoing training		$3,000	$3,000	
Engineering changes	$1,000	$3,000	$4,000	
Service agreement				
Fixed operation and maintenance costs		$1,000	$1,000	
Spares	$2,000		$2,000	
Online testing		$18,000	$18,000	
Repair costs	$1,000	$500	$1,500	
Hazard costs			$12,000	
Spurious trips costs			$4,000	
Annual costs subtotal				$45,500
Present value for annual costs				$567,031
(20 years, 5% discount rate)				
Total Life Cycle Costs				$671,031

Table 14-3   Life Cycle Costs for Case 1

	Material $	Labor $	Total cost $	Subtotal $
Initial Fixed Costs				
Safety classification		$2,000	$2,000	
SRS/design specifications		$5,000	$5,000	
Detailed design and engineering		$30,000	$30,000	
Sensors	$12,000		$12,000	
Final elements	$6,000		$6,000	
Logic system	$30,000		$30,000	
Miscellaneous: power, wiring, junction boxes	$5,000		$5,000	
Initial training		$15,000	$15,000	
FAT/installation/PSAT	$5,000	$20,000	$25,000	
Start-up and correction	$1,000	$2,000	$3,000	
Fixed costs subtotal				$133,000
Annual Costs				
Ongoing training		$3,000	$3,000	
Engineering changes	$1,000	$1,000	$2,000	
Service agreement		$2,000	$2,000	
Fixed operation and maintenance costs		$1,000	$1,000	
Spares	$4,000		$4,000	
Online testing		$18,000	$18,000	
Repair costs	$1,000	$500	$1,500	
Hazard costs				
Spurious trips costs			$4,000	
Annual costs subtotal				$35,500
Present value for annual costs				**$442,408**
(20 years, 5% discount rate)				
Total Life Cycle Costs				**$575,408**

Table 14-4   Life Cycle Costs for Case 2

14.7 Optimizing Safety, Reliability, and Life Cycle Costs

Reliability modeling and life cycle cost analysis are very effective tools for optimizing the design of the safety system. By means of the models and the cost analysis one can better justify an architecture by showing the long-term benefits. Moreover, with these two tools we can do the following:

- Better focus on the specific elements of the safety system that require increased attention. For example, in case 2 it was obvious that the valves remain an issue. Various options can be analyzed based on their long-term impact on reliability and cost.

- A decision can easily be made regarding the selection of the logic system, for example, the use of relay logic versus a redundant programmable system.

- The need for redundancy can be better established.

The importance and benefits of enhanced diagnostics will become apparent if the diagnostic coverage factors are accounted for in the λ_d values that are used (see Section 9.3.2)

The reliability modeling and life cycle cost analysis approach can also be used to analyze whether safety instrumented systems can be used as an alternative to other layers. It is essential that process and mechanical specialists be involved in this analysis.

Summary

Most corporations have a system in place for identifying the need for a safety system. These systems are primarily geared toward satisfying the safety requirements, hence justifying a safety system purely from a safety point of view is not usually an issue.

What is not adequately addressed in safety system justification is the link between the safety, reliability, and the life cycle costs of the safety system. A safety system has to be justified not only for safety but also for its reliability and life cycle cost.

The life cycle model is the ideal tool for justifying the design and installation of a safety system so as to satisfy all the requirements.

References

1. OSHA, *Process Safety Management of Highly Hazardous Chemicals, United States Code of Federal Regulations*, 29 CFR 1910.119 (Washington, DC: Government Printing Office).

2. William M. Goble, *Evaluating Control System Safety and Reliability: Techniques and Applications* (Research Triangle Park, NC: ISA, 1998).

SIS DESIGN CHECKLIST

He tells me to look for dirt daubers.
What the heck are dirt daubers?!

Gruhn

15.1   Introduction

Using a checklist will not in and of itself lead to safer systems. It is by following the procedures outlined in the checklist, which are based on industry standards and accumulated knowledge (some learned "the hard way"), that you will achieve safer systems. The checklist presented in this chapter is an attempt to list as many procedures and common practices as possible in the hope that by following a systematic review of the overall design process nothing will "fall through the cracks" and be forgotten.

This checklist is composed of several sections, each corresponding to different portions of the safety life cycle as described in the ANSI/ISA S84.01-1996 standard.[1] Different sections of the checklist are intended for different groups involved in the overall system design, from the user and contractor to the vendor and system integrator. Exactly who has what responsibility may vary from project to project. The checklist, therefore, does not dictate who has which responsibilities; it only summarizes items in the various life cycle steps.

Why bother? As noted in previous chapters, the English Health and Safety Executive (HSE) reviewed and published the findings of thirty-four accidents that were the direct result of control and safety system failures.[2] (Its findings are summarized in Figure 15-1.) Forty-four percent of the accidents were due to *incorrect specifications* (functional and integrity), and 20 percent were due to changes made after commissioning. One can easily see that the majority of these issues focus on *user* activities. (Hindsight is always twenty-twenty; foresight is a bit more difficult.) Industry standards, and so this checklist, attempt to cover *all* of the issues and not just focus on any one particular area.

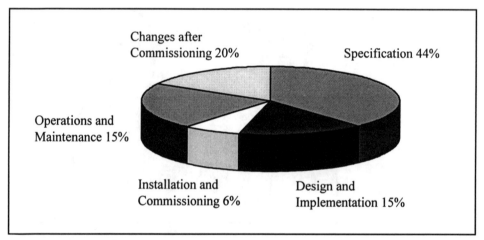

Figure 15-1   Primary Accident Cause by Design Phase As Found by the United Kingdom's HSE

The checklist in this chapter should in no way be considered final or complete. As others review and use it, they are encouraged to add to it (for the benefit of others who may use it in the future). In fact, each section starts with a new page, leaving ample space for your additions and suggestions.

15.2   Overview

The checklist outlined in this chapter covers the following steps:

1.   safety requirements specification (SRS)
2.   conceptual SIS design
3.   detailed SIS design
4.   power and grounding
5.   field devices
6.   operator interface
7.   maintenance/engineering interface
8.   communications
9.   hardware specification
10. hardware manufacture
11. application logic requirements
12. embedded (vendor) software
13. software coding
14. factory test
15. installation and commissioning
16. operations and maintenance
17. testing
18. management of change
19. decommissioning

15.3   Section 1: Safety Requirements Specification

Item #	Item	Circle a choice			Comments
1.1	Do the safety requirements originate from a systematic hazard assessment? If not, what are the requirements based on?	Y	N	N/A	
1.2	Is there a clear and concise description of each safety-related function to be implemented in the SIS?	Y	N	N/A	
1.3	Has the safe state of the process been defined for each operating state of the plant? (e.g., start-up, normal operation, maintenance, etc.)	Y	N	N/A	
1.4	Are safety functions defined for each operating state of the plant?	Y	N	N/A	
1.5	Are performance requirements (e.g., speed, accuracy, etc.) defined for each safety function?	Y	N	N/A	
1.6	Are sensor inputs defined with regard to range, accuracy, noise limits, bandwidth, etc.?	Y	N	N/A	
1.7	Are outputs defined with regard to range, accuracy, update frequency, etc.?	Y	N	N/A	
1.8	In the event of system failure, are sufficient information and means available for the operator to assume safe control?	Y	N	N/A	
1.9	Is the operator interface defined in terms of data display, alarms, etc.?	Y	N	N/A	
1.10	Have local or application-specific regulatory requirements been considered?	Y	N	N/A	
1.11	Has the operation and implementation of resets been defined for each input and output?	Y	N	N/A	
1.12	Have process common cause considerations (e.g., corrosion, plugging, coating, etc.) been considered?	Y	N	N/A	
1.13	Has the required safety integrity level (SIL) been determined for each safety function?	Y	N	N/A	

15.4   Section 2: Conceptual SIS Design

Item #	Item	Circle a choice			Comments
2.1	Are safety functions being handled by a system that is completely separate from the process control? If not, what is the justification?	Y	N	N/A	
2.2	If multiple functions are being performed within the same logic solver, do the shared components meet the highest SIL requirement?	Y	N	N/A	
2.3	Has the technology and level of redundancy been selected for each safety function? If so, what is it?	Y	N	N/A	
2.4	Has the SIL for each safety function been analyzed and documented in a quantitative manner? If not, what is the justification for the system configuration?	Y	N	N/A	

15.5   Section 3: Detailed SIS Design

Item #	Item	Circle a choice			Comments
3.1	Are design documents under the control of a formal revision and release program?	Y	N	N/A	
3.2	Has the SIL of the final system been analyzed and documented in a quantitative manner? If not, what is the justification for the system configuration?	Y	N	N/A	
3.3	Are suitable interfaces between field devices and the logic solver defined?	Y	N	N/A	
3.4	Are suitable communications interfaces defined in terms of protocols and the information to be exchanged?	Y	N	N/A	
3.5	Are there provisions for future expansion?	Y	N	N/A	
3.6	Are there provisions for incorporating changes as the design proceeds?	Y	N	N/A	
3.7	Is the system fail-safe in terms of:				
	a) Loss of power?	Y	N	N/A	
	b) Loss of instrument air?	Y	N	N/A	
	c) Field cable faults?	Y	N	N/A	
3.8	Can the action of a nonsafety function interrupt or compromise any safety functions?	Y	N	N/A	
3.9	Are the safe states of each system component defined?	Y	N	N/A	
3.10	Has the impact of the failure of each component in the system been considered and the action that must be taken defined?	Y	N	N/A	
3.11	Is field I/O power separate from other circuits?	Y	N	N/A	
3.12	Are I/O bypasses incorporated?	Y	N	N/A	
3.13	When an input bypass is enabled can the state of the sensor still be determined?	Y	N	N/A	
3.14	Are there means for alarming a bypass after a predetermined time interval?	Y	N	N/A	
3.15	Does the system incorporate manual resetting to restart production? If not, what is the justification?	Y	N	N/A	

15.6   Section 4: Power and Grounding

Item #	Item	Circle a choice			Comments
4.1	Are power supplies DC? If not, what is the justification?	Y	N	N/A	
4.2	Is a redundant power source available? If not, what is the justification?	Y	N	N/A	
4.3	Has the impact of power failures been considered?	Y	N	N/A	
4.4	Have the following power concerns been addressed:				
	a) Voltage and current range, including inrush current?	Y	N	N/A	
	b) Frequency range?	Y	N	N/A	
	c) Harmonics?	Y	N	N/A	
	d) Nonlinear loads?	Y	N	N/A	
	e) AC transfer time?	Y	N	N/A	
	f) Overload and short-circuit protection?	Y	N	N/A	
	g) Lightning protection?	Y	N	N/A	
	h) Protection against transient spikes, surges, brownouts, and noise?	Y	N	N/A	
	i) Under- and overvoltage?	Y	N	N/A	
4.5	Have the following grounding concerns been addressed:				
	a) Corrosion protection?	Y	N	N/A	
	b) Cathodic protection?	Y	N	N/A	
	c) Static electricity protection?	Y	N	N/A	
	d) Shield ground?	Y	N	N/A	
	e) Test ground?	Y	N	N/A	
	f) Intrinsic safety barrier grounds?	Y	N	N/A	

15.7   Section 5: Field Devices

Item #	Item	Circle a choice			Comments
5.1	Have vendors provided failure rate, failure mode, and diagnostic coverage information?	Y	N	N/A	
5.2	Have vendors provided recommended manual test intervals and procedures?	Y	N	N/A	
5.3	Will means be available for periodically checking the devices for hidden dangerous failures?	Y	N	N/A	
5.4	Are circuits normally energized? If not, is line monitoring (supervised circuits) being incorporated?	Y	N	N/A	
5.5	Does each device have its own dedicated wiring? If not, what is the justification?	Y	N	N/A	
5.6	If smart sensors are being used, are they write protected?	Y	N	N/A	
5.7	Have minimum, as well as maximum, electrical loads been considered for field I/O circuits?	Y	N	N/A	
5.8	Is feedback available to tell if the final element has moved to its commanded state?	Y	N	N/A	
5.9	Have materials (seals, etc.) been properly selected for the particular application?	Y	N	N/A	
5.10	Does the user have good field experience with the devices in other applications?	Y	N	N/A	
5.11	Are solenoid valves protected from plugging, dirt, insects, freezing, etc.? What measures have been applied?	Y	N	N/A	
5.12	Have the following areas been considered for final elements:				
	a) Opening and closing speeds?	Y	N	N/A	
	b) Shutoff differential pressure?	Y	N	N/A	
	c) Leakage?	Y	N	N/A	
	d) Fire resistance of body, actuator, and impulse lines?	Y	N	N/A	
5.13	Are safety-critical field devices identified in some unique manner (e.g., color coding)?	Y	N	N/A	

15.8   Section 6: Operator Interface

Item #	Item	Circle a choice			Comments
6.1	Has failure (loss) of the interface been considered?	Y	N	N/A	
6.2	Are alternate means available to bring the process to a safe state?	Y	N	N/A	
6.3	Are the following shown on the interface:				
	a) Where the process is in a sequence?	Y	N	N/A	
	b) Indication that a SIS action has occurred?	Y	N	N/A	
	c) Indication that a SIS function is bypassed?	Y	N	N/A	
	d) Indication that a SIS component or subsystem has failed or is in a degraded state?	Y	N	N/A	
	e) Status of field devices?	Y	N	N/A	
6.4	Is the update time appropriate for the application under emergency conditions?	Y	N	N/A	
6.5	Have the operators been checked for color blindness?	Y	N	N/A	
6.6	Is it possible to change SIS program logic from the operator interface?	Y	N	N/A	
6.7	Do parameters that can be changed have security access protection?	Y	N	N/A	

15.9   Section 7: Maintenance/Engineering Interface

Item #	Item	Circle a choice			Comments
7.1	Can the failure of this interface adversely affect the SIS?	Y	N	N/A	
7.2	Is there adequate access security? What methods are utilized?	Y	N	N/A	

15.10 Section 8: Communications

Item #	Item	Circle a choice			Comments
8.1	Can communication failures have an adverse affect on the SIS?	Y	N	N/A	
8.2	Are communication signals isolated from other energy sources?	Y	N	N/A	
8.3	Has write protection been implemented so that external systems cannot corrupt SIS memory? If not, why?	Y	N	N/A	

15.11   Section 9: Hardware Specification

Item #	Item	Circle a choice			Comments
9.1	Has the physical operating environment been defined and have suitable specifications been set for:				
	a) Temperature range?	Y	N	N/A	
	b) Humidity?	Y	N	N/A	
	c) Vibration and shock?	Y	N	N/A	
	d) Ingress of dust and/or water?	Y	N	N/A	
	e) Contaminating gases?	Y	N	N/A	
	f) Hazardous atmospheres?	Y	N	N/A	
	g) Power supply voltage tolerance?	Y	N	N/A	
	h) Power supply interruptions?	Y	N	N/A	
	i) Electrical interference?				
	j) Ionizing radiation?	Y	N	N/A	
9.2	Are the failure modes known for all components?	Y	N	N/A	
9.3	Has the vendor supplied quantitative safe and dangerous system failure rates, including the assumptions and component data used?	Y	N	N/A	
9.4	Has the vendor provided diagnostic coverage values for its system? If so, what are they?	Y	N	N/A	
9.5	Are logic system components (I/O modules, CPU, communication modules, etc.) all from the same vendor? If not, what is the justification?	Y	N	N/A	
9.6	Has the resulting action of restoring power to the system been considered?	Y	N	N/A	
9.7	Are all I/O modules protected from voltage spikes?	Y	N	N/A	
9.8	If redundant devices or systems are being considered, have measures been taken to minimize potential common cause problems, and if so, what are they?	Y	N	N/A	

15.12 Section 10: Hardware Manufacture

Item #	Item	Circle a choice			Comments
10.1	Can the vendor provide evidence of an independent safety assessment of the hardware?	Y	N	N/A	
10.2	Does the vendor maintain a formal revision and release control program?	Y	N	N/A	
10.3	Are there visible indications of version numbers on the hardware?	Y	N	N/A	
10.4	Does the vendor have specifications and procedures for the quality of materials, workmanship, and inspections?	Y	N	N/A	
10.5	Are adequate precautions taken to prevent damage due to static discharge?	Y	N	N/A	

15.13 Section 11: Application Logic Requirements

Item #	Item	Circle a choice			Comments
11.1	Do all parties have a formal revision and release control program for application logic?	Y	N	N/A	
11.2	Is the logic written in a clear and unambiguous manner that is understandable to all parties?	Y	N	N/A	
11.3	Does the program include comments?	Y	N	N/A	
11.4	Within the logic specification, is there a clear and concise statement of:				
	a) Each safety-related function?	Y	N	N/A	
	b) The information to be given to the operator?	Y	N	N/A	
	c) The required action of each operator command, including illegal or unexpected commands?	Y	N	N/A	
	d) The communication requirements between the SIS and other equipment?	Y	N	N/A	
	e) The initial states for all internal variables and external interfaces?	Y	N	N/A	
	f) The required action for out-of-range variables?	Y	N	N/A	

15.14 Section 12: Embedded (Vendor) Software

Item #	Item	Circle a choice			Comments
12.1	Can the vendor provide evidence of an independent safety assessment of all embedded software?	Y	N	N/A	
12.2	Has the software been in satisfactory use in similar applications for a significant period of time?	Y	N	N/A	
12.3	Is the vendor software documented sufficiently for the user to understand its operation and how to implement the desired functionality?	Y	N	N/A	
12.4	Are the results of abnormal math operations fully documented?	Y	N	N/A	
12.5	Are there procedures for controlling the software versions in use and updating all similar systems?	Y	N	N/A	
12.6	For spares that contain firmware, is there a procedure for ensuring that all modules are compatible?	Y	N	N/A	
12.7	Can software versions in use easily be checked?	Y	N	N/A	
12.8	If errors are found in embedded software, are they reported to and corrected by the vendor and incorporated into the SIS only after checking and testing the corrected code?	Y	N	N/A	
12.9	Does the manufacturer provide competent technical support?	Y	N	N/A	

15.15 Section 13: Software Coding

Item #	Item	Circle a choice			Comments
13.1	Are there standards and procedures for software coding?	Y	N	N/A	
13.2	Is there a procedure for documenting and correcting any deficiencies in the specification or design revealed during the coding phase?	Y	N	N/A	
13.3	Are departures from or enhancements to the requirements of the design documented?	Y	N	N/A	
13.4	Is a formal language or some other means taken to assure that the program is both precise and unambiguous?	Y	N	N/A	
13.5	Is there a procedure for generating and maintaining adequate documentation?	Y	N	N/A	
13.6	Does the programming language encourage the use of small and manageable modules?	Y	N	N/A	
13.7	Does the code include adequate comments?	Y	N	N/A	
13.8	Are design reviews carried out during program development that involve users, designers, and programmers?	Y	N	N/A	
13.9	Does the software contain adequate error detection facilities associated with error containment, recovery, or safe shutdown?	Y	N	N/A	
13.10	Are all functions testable?	Y	N	N/A	
13.11	Is the final code checked against the requirements by persons other than those producing the code?	Y	N	N/A	
13.12	Is a well-established compiler/assembler used?	Y	N	N/A	
13.13	Is the compiler/assembler certified to recognized standards?	Y	N	N/A	

15.16 Section 14: Factory Test

Item #	Item	Circle a choice			Comments
14.1	Are there standards and procedures for testing the finished system?	Y	N	N/A	
14.2	Are records maintained outlining the tests to be carried out and the test results?	Y	N	N/A	
14.3	Is there a procedure for documenting and correcting any deficiencies in the specification, design, or programming revealed during testing?	Y	N	N/A	
14.4	Is testing carried out by persons other than those producing the code?	Y	N	N/A	
14.5	Is software tested in the target system rather than simulated?	Y	N	N/A	
14.6	Is each control flow or logic path tested?	Y	N	N/A	
14.7	Have arithmetic functions been tested with minimum and maximum values to ensure that no overflow conditions are reached?	Y	N	N/A	
14.8	Are there tests to simulate exceptions as well as normal conditions?	Y	N	N/A	

15.17 Section 15: Installation and Commissioning

Item #	Item	Circle a choice			Comments
15.1	Have personnel received appropriate training?	Y	N	N/A	
15.2	Is there sufficient independence between those carrying out the work and those inspecting it?	Y	N	N/A	
15.3	Have adequate precautions been taken for storing items during installation?	Y	N	N/A	
15.4	Are installation procedures for all devices sufficient in detail so as not to leave important interpretations or decisions to installation personnel?	Y	N	N/A	
15.5	Has the SIS been inspected for any damage caused during installation?	Y	N	N/A	
15.6	Are items such as cabinets, junction boxes, and cables protected from:				
	a) Steam leaks?	Y	N	N/A	
	b) Water leaks?	Y	N	N/A	
	c) Oil leaks?	Y	N	N/A	
	d) Heat sources?	Y	N	N/A	
	e) Mechanical damage?	Y	N	N/A	
	f) Corrosion (e.g., process fluids flowing from damaged sensors to junction boxes, the logic cabinet, or the control room)?	Y	N	N/A	
	g) Combustible atmospheres?	Y	N	N/A	
15.7	Are safety-related systems clearly identified so as to prevent inadvertent tampering?	Y	N	N/A	
15.8	Has the proper operation of the following items been confirmed:				
	a) Equipment and wiring installed properly?	Y	N	N/A	
	b) Energy sources are operational?	Y	N	N/A	
	c) All field devices have been calibrated?	Y	N	N/A	
	d) All field devices are operational?	Y	N	N/A	
	e) The logic solver is operational?	Y	N	N/A	
	f) Communication with other systems?	Y	N	N/A	
	g) Operation and indication of bypasses?	Y	N	N/A	
	h) Operation of resets?	Y	N	N/A	
	i) Operation of manual shutdowns?	Y	N	N/A	
15.9	Is the documentation consistent with the actual installation?	Y	N	N/A	
15.10	Is there documentation showing the following:				
	a) Identification of the system being commissioned?	Y	N	N/A	
	b) Confirmation that commissioning has been successfully completed?	Y	N	N/A	
	c) The date the system was commissioned?	Y	N	N/A	
	d) The procedures used to commission the system?	Y	N	N/A	
	e) The authorized signature indicating the system was successfully commissioned?	Y	N	N/A	

15.18 Section 16: Operations and Maintenance

Item #	Item	Circle a choice			Comments
16.1	Have employees been adequately trained on the operating and maintenance procedures for the system?	Y	N	N/A	
16.2	Are operating procedures adequately documented?	Y	N	N/A	
16.3	Is there a user/operator/maintenance manual for the system?	Y	N	N/A	
16.4	Does the manual describe:				
	a) The limits of safe operation and the implications of exceeding them?	Y	N	N/A	
	b) How the system takes the process to a safe state?	Y	N	N/A	
	c) The risk associated with system failures and the actions required for different failures?	Y	N	N/A	
16.5	Are there means for limiting access only to authorized personnel?	Y	N	N/A	
16.6	Can all operational settings be readily inspected to ensure they are correct at all times?	Y	N	N/A	
16.7	Are there means for limiting the range of input trip settings?	Y	N	N/A	
16.8	Have adequate means been established for bypassing safety functions?	Y	N	N/A	
16.9	When functions are bypassed are they clearly indicated?	Y	N	N/A	
16.10	Have documented procedures been established to control the application and removal of bypasses?	Y	N	N/A	
16.11	Have documented procedures been established to ensure the safety of the plant during SIS maintenance?	Y	N	N/A	
16.12	Are maintenance procedures sufficient in detail so as not to leave important interpretations or decisions to maintenance personnel?	Y	N	N/A	
16.13	Are maintenance activities and schedules defined for all portions of the system?	Y	N	N/A	
16.14	Are procedures periodically reviewed?	Y	N	N/A	
16.15	Are procedures in place to prevent unauthorized tampering?	Y	N	N/A	
16.16	Are there means for verifying that repair is carried out in a time consistent with that assumed in the safety assessment?	Y	N	N/A	
16.17	Are maintenance and operational procedures in place to minimize the introduction of potential common cause problems?	Y	N	N/A	
16.18	Is the documentation consistent with the actual maintenance and operating procedures?	Y	N	N/A	

15.19 Section 17: Testing

Item #	Item	Circle a choice			Comments
17.1	Have provisions been designed to allow proof testing of all safety functions, including field devices, and are they documented?	Y	N	N/A	
17.2	Are test procedures sufficient in detail so as not to leave important interpretations or decisions to maintenance personnel?	Y	N	N/A	
17.3	Has the basis for the periodic test interval been documented?	Y	N	N/A	
17.4	Are the following items tested:				
	a) Impulse lines?	Y	N	N/A	
	b) Sensing devices?	Y	N	N/A	
	c) Application logic, computations, and/or sequences?	Y	N	N/A	
	d) Trip points?	Y	N	N/A	
	e) Alarm functions?	Y	N	N/A	
	f) Speed of response?	Y	N	N/A	
	g) Final elements?	Y	N	N/A	
	h) Manual trips?	Y	N	N/A	
	i) Diagnostics?	Y	N	N/A	
17.5	Is there a fault-reporting system?	Y	N	N/A	
17.6	Are procedures in place to compare actual performance against the predicted or required performance?	Y	N	N/A	
17.7	Are there documented procedures for correcting any deficiencies found?	Y	N	N/A	
17.8	Is the calibration of test equipment verified?	Y	N	N/A	
17.9	Are test records maintained?	Y	N	N/A	
17.10	Do test records show:				
	a) Date of inspection/test?	Y	N	N/A	
	b) Name of person conducting the test?	Y	N	N/A	
	c) Identification of the device being tested?	Y	N	N/A	
	d) Results of the test?	Y	N	N/A	
17.11	Are testing procedures in place to minimize the introduction of potential common cause problems?	Y	N	N/A	

15.20   Section 18: Management of Change

Item #	Item	Circle a choice			Comments
18.1	Are there approval procedures that consider the safety implications of all modifications, such as:				
	a) The technical basis for the change?	Y	N	N/A	
	b) The impact on safety and health?	Y	N	N/A	
	c) The impact on operating/maintenance procedures?	Y	N	N/A	
	d) The time required?	Y	N	N/A	
	e) The effect on response time?	Y	N	N/A	
18.2	Have all affected departments been appraised of the change?	Y	N	N/A	
18.3	Has the proposed change initiated a return to the appropriate phase of the life cycle?	Y	N	N/A	
18.4	Has the project documentation (e.g., operating, test, maintenance procedures, etc.) been altered to reflect the change?	Y	N	N/A	
18.5	Has the complete system been tested after changes have been introduced and the results documented?	Y	N	N/A	
18.6	Are there documented procedures to verify that changes have been satisfactorily completed?	Y	N	N/A	
18.7	Is access to the hardware and software limited to authorized and competent personnel?	Y	N	N/A	
18.8	Is access to the project documentation limited to authorized personnel?	Y	N	N/A	
18.9	Are project documents subject to appropriate revision control?	Y	N	N/A	
18.10	Have the consequences of incorporating new versions of software been considered?	Y	N	N/A	

15.21   Section 19: Decommissioning

Item #	Item	Circle a choice			Comments
19.1	Have management-of-change procedures been followed for decommissioning activities?	Y	N	N/A	
19.2	Has the impact on adjacent operating units and facilities been evaluated?	Y	N	N/A	

References

1. ANSI/ISA, *Application of Safety Instrumented System for the Process Industries*, ANSI/ISA S84.01-1996 (Research Triangle Park, NC: ISA, 1996), ISBN 1-55617-590-6.

2. U.K. Health and Safety Executive, *Programmable Electronic Systems in Safety-Related Applications, Part 2: General Technical Guidelines* (Sheffield, UK: U.K. HSE, 1987), ISBN 011-883906-3.

3. American Institute of Chemical Engineers, *Guidelines for Safe Automation of Chemical Processes* (New York: AIChE, Center for Chemical Process Safety, 1993), ISBN 0-8169-0554-1.

4. U.K. Health and Safety Executive, *Out of Control: Why Control Systems Go Wrong and How to Prevent Failure* (Sheffield, UK: U.K. HSE, 1995), ISBN 0-7176-0847-6.

CASE STUDY

16.1 Introduction

The case study presented in this chapter is intended to show how the material and techniques outlined in Chapters 1 to 15 can be used to specify, design, install, commission, and maintain a typical safety instrumented system (SIS). Our intention is to follow the life cycle model presented in Chapter 1 so as to better focus on the approach being used in the case study. By reviewing the solutions reached to the specific issues in this case study the reader will be able to further clarify the techniques, better understand and appreciate the need for good documentation, and resolve possible misunderstandings. This chapter should serve as an additional guide to those involved in specifying and designing safety systems.

Since the emphasis in this chapter is on the approach and not the problem or the final solution, the case studied here has been simplified so that the focus remains on "how to" rather than "how many."

A Word of Caution

The controls and protective instrumentation discussed in this chapter are oversimplifications offered for study purposes only. They do not in any way reflect what is or what may be required for actual installations. Also, the solutions and the system proposed do not necessarily comply with some internationally recognized standards, for example, those of the National Fire Protection Association.

16.2 Case Description: Furnace/Fired Heater Safety Shutdown System

The safety instrumented system for the fuel controls of a furnace/fired heater, which is part of a crude unit in an oil refinery, has to be upgraded because of the plant management's concern that the existing shutdown (S/D) and safety systems are inadequate, ineffective, and unreliable. The furnace is natural draft and natural gas fired, with continuous pilots.

Figure 16-1 shows a diagram of a furnace and the basic process controls associated with the fuels.

16.2.1   Description of Basic Process Controls

The feed flow to the furnace is controlled by regulatory control loops FC-9 and FC-10.

A local self-regulating pressure controller PCV-6 controls the pilot gas pressure. The fuel gas flow and coil outlet temperature controls are part of a cascade control system. The gas flow controller FC-3 is in cascade with the temperature controller TC-8. The flow controller manipulates valve FV-3 to control the gas flow at a set point manipulated by the temperature controller.

All regulatory control loops are in a distributed control system (DCS), which is located in a control room approximately two hundred feet away from the furnace. As a result of a hazard review for the furnaces the hazards shown in Table 16-1 were identified.

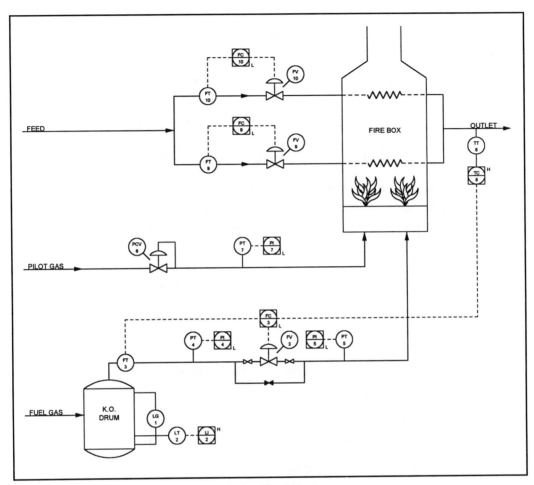

Figure 16-1   Basic Process Controls for Fired Heater

Hazard	Possible Cause	Consequence	Likelihood
Furnace explosion	Loss of pilots.	Loss of life	Medium
Furnace fire	Loss of pilots. Loss of flow in furnace tube.	$1M loss	Medium
Tube failure	Loss of flow in furnace tube.	$1M loss	Medium

Table 16-1   Furnace Hazards

It was felt that no additional noninstrumented layers could be applied to mitigate the events, so the following safety and protective instrumentation was recommended.

1. Fuel gas to be shut off on low pass flow on any pass (with twenty-second delay).

2. The fuel gas and pilots to be shut off on low pilot gas pressure.

3. A dedicated hardwired switch to be provided in the main control room for the manual shutoff of fuel and pilots.

A demand on the safety system is likely to occur every six months. Also, one nuisance trip per year of the furnace caused by a failure of the safety system can be tolerated. The cost impact of a nuisance trip is approximately $20,000.

16.3 Scope of Analysis

The scope of the analysis includes the following safety life cycle steps:

1. Define target SIL

2. Develop safety requirement specification (SRS)

3. Perform SIS conceptual design

4. Life cycle costs

5. Verify that the SIS conceptual design meets the SIL

6. Perform detailed design

7. Install the SIS, commission it, and perform pre-start-up tests

8. Establish operation and maintenance procedures

16.4 Define Target SIL

Four methods for determining the SIL were outlined in Chapter 6:

Qualitative: Using the safety layer matrix

Qualitative: Using the IEC risk model

Quantitative: Using mean-time-between-hazards (MTTR) and demand rate calculations

Quantitative: Layers of protection analysis (LOPA)

Note: A SIL has to be determined for each safety function.

Since the HAZOP data in Table 16-1 provided the consequences and likelihood of the hazardous events, the qualitative risk matrix will be used. There are no additional protective layers. Table 16-1 can therefore be simplified for this application using the risk matrix below:

		L	M	H
	3	2	3	3
Consequence	2	2	2	3
	1	1	2	2
		L	M	H

Likelihood

The SIL based on this matrix is shown in Table 16-2.

Item	Safety Function	Consequence		Likelihood		SIL
1	Low or loss of flow in passes	Tube failure $1M loss	2	Medium	M	2
		Furnace fire $1M loss	2	Medium	M	
2	Low pilot gas pressure	Explosion Loss of life	3	Medium	M	3

Table 16-2   Safety Integrity Levels for Fired Heater

16.5 Develop Safety Requirement Specification (SRS)

Chapter 5 detailed the documentation required for the safety requirement specification (SRS), which consists of the functional specification (what the system does) and the integrity specification (how well it does it). The SRS summary table shown in Table 16-3 summarizes the key information that should be part of the SRS and is used to document it. By listing all the requirements in a single table you can easily cross-check what should be provided.

16.5.1 Safety Requirement Specification—Summary Table

The "Details of Requirement" column in Table 16-3 identifies whether the information pertaining to each item should be provided for this application (such as documentation and input). If it is required, any special comments are included in this column.

Item	Details of Requirement
Documentation and Input Requirements	
P&IDs	Required. Figure 16-2 is a simplification of the actual P&ID.
Cause-and-effect diagram	See Table 16-4.
Logic diagrams	The cause-and-effect drawing is adequate for identifying the logic requirements.
Process data sheets	Have to be provided for all the field devices, i.e., FT-22, FT-23, PT-7, PT-24, and the four trip valves.
Process information (incident cause, dynamics, final elements, etc.) of each potential hazardous event that requires an SIS.	A detailed description is required of the details as to how explosions, fires, or tube failures can occur and how the protective instruments will mitigate these occurrences. Speed of response and accuracy of the protective system. Special requirements for the trip valves, i.e., fire safety and tight shutoff.
Process common cause failure considerations such as corrosion, plugging, coating, etc.	The crude flow measurement is a difficult application due to high viscosity and freeze-up. An in-line device is preferable. For this application, the site has had good experience with Coriolis meters. Corrosion due to hydrogen sulfide in the atmosphere has to be addressed.
Regulatory requirements impacting the SIS.	NFPA standard 8501 can be used as a reference. Full compliance with the standard is not mandatory.
Other	
Functional Requirements	
The definition of the safe state of the process for each of the identified events	The safe state is to shut off the fuel gas, pilot gas, and the feed to the furnace.
The process inputs to the SIS and its trip points.	See Table 16-4.
The normal operating range of the process variables and their operating limits.	See Table 16-4.
The process outputs from the SIS and their actions.	See Table 16-4.
The functional relationship between process inputs and outputs, including logic, math functions, and any required permissives.	See Table 16-4.
Selection of deenergized to trip or energized to trip.	The complete safety system has to be deenergized to trip.
Consideration for manual shutdown.	One hardwired shutdown switch must be located in the main control room (HS-21).
Action(s) to be taken on the loss of energy source(s) to the SIS.	All trip valves to close fully.
Response time requirement for the SIS to bring the process to a safe state.	5 seconds is acceptable.
Operators' response action to any overt fault.	Immediate response. Maintenance must be assigned highest priority. While problem is being rectified, closer monitoring of critical parameters must be put in place by operations.
Human-machine interface requirements.	Dedicated hardwired alarms in the main control room are required for any safety system fault or trip condition. Manual S/D button required.
Reset function(s).	The trip valves have to be provided with manual reset solenoids.

Safety Integrity Requirements	
A list of the safety function(s) required and the SIL of each safety function.	See Table 16-4.
Requirements for diagnostics to achieve the required SIL.	Smart field sensors are to be provided to utilize their diagnostic capabilities. The sensor signals are to go to maximum value if a failure is diagnosed. This state is to be alarmed on the hardwired annunciator. The trip valves are to be provided with limit switches to verify that they have closed when requested by the logic system.
Requirements for maintenance and testing to achieve the required SIL.	The online testing frequency is as described in Section 16.11.
Reliability requirements if spurious trips may be hazardous.	The spurious trip rate must be calculated for compliance with the HAZOP requirements.

Table 16-3   Summary of Key Information Required for SRS

TAG #	DESCRIPTION	SIL	INSTRUMENT RANGE	TRIP VALUE	UNITS	CLOSES VALVE XV-30A	CLOSES VALVE XV-30B	CLOSES VALVE XV-31A	CLOSES VALVE XV-31B	NOTES
FT-22	Pass flow A to furnace	2	0-500	100.0	B/H	X	X			1
FT-23	Pass flow B to furnace	2	0-500	100.0	B/H	X	X			1
PT-7/24	Pilot gas pressure	3	0-30	5	PSIG	X	X	X	X	2
	Loss of control power					X	X	X	X	
	Loss of instrument air					X	X	X	X	
HS-21	Manual shutdown					X	X	X	X	

Notes:
1. Twenty-second delay required before closing valves.
2. Solenoids with manual reset to be provided for all four trip valves.

Table 16-4   Cause-and-Effect Diagram (See Figure 16-2)

16.6 Perform SIS Conceptual Design

The conceptual design must comply with the corporate standards that govern SIS design. Table 16-5 is a summary of the corporate guidelines relating to SIL selection.

SIL	Sensors	Logic Solver	Final Elements
3	Sensor redundancy required either 1oo2 or 2oo3, depending on spurious trip requirements.	Redundant PES required.	1oo2 voting required.
2	Redundancy may or may not be required. The initial option is not to have redundancy. Select redundancy if warranted by PFD_{avg} calculations.	Redundant PES required.	Redundancy may or may not be required. Initial option is not to have redundancy. Select redundancy if warranted by PFD_{avg} calculations.
1	Single sensor.	Single PES or relay logic.	Single device.

Table 16-5   **SIS Design Guidelines based on SIL Levels**

Based on the criteria in Table 16-5, the proposed system is as shown in Figure 16-2.

16.6.1 Conceptual Design Requirements

The conceptual design (see Chapter 10) builds on and supplements the safety requirement specification (SRS) in Section 16.5 as well as the corporate design guidelines. The key information required by the engineering contractor to complete the detailed engineering package is provided in the conceptual design. The design must adhere to the company standards and procedures. Also, there must be no contradiction between the SRS and the conceptual design requirements.

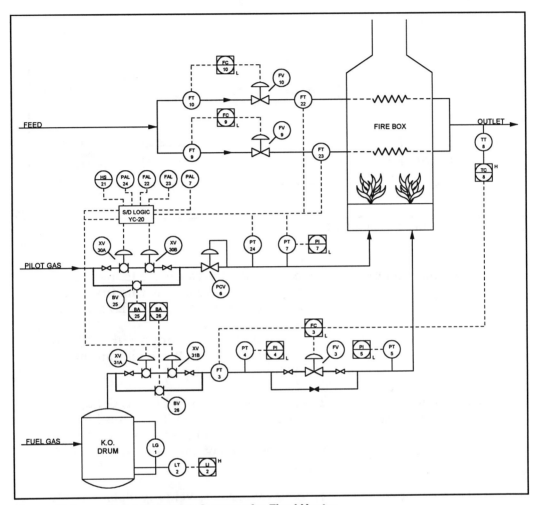

Figure 16-2   Safety Interlock Systems for Fired Heater

The following list shows the basic conceptual design requirements.

System architecture: Based on the corporate guidelines, the PES shall be redundant. The corporate standard for redundant PES is to be selected. Locate the system in the main control building. 1oo2 voting is required for the pilot gas pressure transmitters' signals and for the fuel gas and pilot gas trip valves. Verification that this design meets the SIL requirements is required. Figure 16-3 shows a sketch of the architecture.

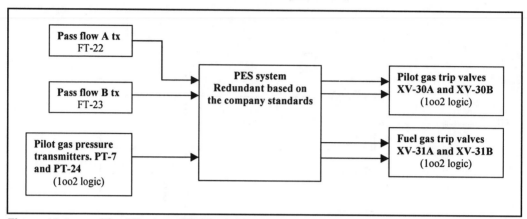

Figure 16-3   Fired Heater - SIS Block Diagram

Minimize common cause:	To reduce common cause, the following specific actions must be taken: (1) wiring from the 1oo2 valves and transmitters to the PES must be segregated from the BPCS wiring, (2) separate uninterruptible power supply (UPS) devices must power the PES, (3) and all transmitters must have separate taps.
Environmental considerations:	The area classification is Class 1, Group D, Division 2. Hydrogen sulfide gas is in the environment around the furnace. During the winter, the ambient temperature can fall to –35°C.
Power supplies:	110-volt, 60-hertz power is available from two separate UPS systems located in the main control room.
Grounding:	Ensure that the grounding techniques for the instrument and power systems are followed.
Bypasses:	For online testing, bypass valves must be installed around the pair of trip valves for the pilot gas and fuel gas trip valves. There must be an alarm in the DCS to indicate that a bypass valve has opened. No other bypasses are required.
Application software:	Ladder logic must be used for all programming in the PES.

Security: The existing corporate security requirements for
                     gaining access to and modifying the PES logic
                     have to be followed.

Operators' interface: The shutdown and diagnostic alarms must be
                     wired to the existing hardwired annunciator.
                     The bypass alarms must be connected to the
                     DCS system.

16.7 Life Cycle Costs

The life cycle costs based on the analysis techniques outlined in Chapter
14, section 14.4 are as follows:

Life Cycle Costs (20 years)	Material $	Labor $	Total cost $	Subtotal $
Initial Fixed Costs				
Safety classification		$1,000	$1,000	
SRS/design specifications		$3,000	$3,000	
Detailed design and engineering		$20,000	$20,000	
Sensors	$24,000		$24,000	
Final elements	$6,000		$6,000	
Logic system	$30,000		$30,000	
Miscellaneous: power, wiring, junction boxes	$4,000		$4,000	
Initial training		$5,000	$5,000	
FAT/installation/PSAT	$4,000	$16,000	$20,000	
Start-up and correction	$1,000	$2,000	$3,000	
Fixed costs subtotal				$116,000
Annual Costs				
Ongoing training		$1,000	$1,000	
Engineering changes	$1,000	$1,000	$2,000	
Service agreement		$1,000	$1,000	
Fixed operation and maintenance costs		$1,000	$1,000	
Spares	$4,000		$4,000	
Online testing		$8000	$8,000	
Repair costs	$1,000	$500	$1,500	
Hazard costs				
Spurious trips costs			$8,000	
Annual costs subtotal				$26,500
Present value for annual costs (20 years, 5% discount rate)				$330,249
Total Life Cycle Costs				$446,249

Table 16-6   Life Cycle Cost Analysis

16.8 Verify That the SIS Conceptual Design Meets the SIL

We have to verify that each trip function meets the SIL requirements. In this case, we have three trips, that is, low pass flow A, low pass flow B, and low pilot gas pressure. The low pilot gas pressure requires SIL 3. We need to verify that this is correct. The same verification procedure can be used for the pass flow trips.

The equations outlined in Chapter 8 will be used for the calculations. A block diagram is shown in Figure 16-4.

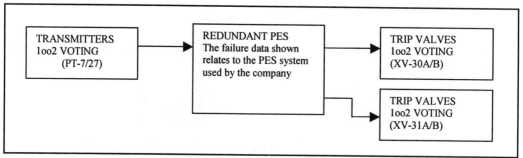

Figure 16-4   Low Pilot Gas Pressure Trip System Interconnection

The following assumptions apply:

>   MTTR: 8 hours
>
>   Average demand rate: Every 6 months
>
>   Online test interval is: 3 months
>
>   Common cause effect is negligible, i.e., β factor is very low.
>
>   The fail-to-danger rate includes all failures and diagnostic coverage.

The fail-safe rate for all transmitters is the same.

Failure data:

Item	Fail-to-Danger Rate λ_d	Fail-to-Safe Rate λ_s
Transmitter PT-7/24	0.01	0.02
Valve and sol^n XV-30A/B XV-31A/B	0.02	0.1
PES	See note 1 below	.01

Note:
1. The PFD_{avg} values for the redundant PES is supplied by the vendor and is . 00001.

Table 16-7   Failure Data for SIS Components

PFD_{avg} calculations: (See Chapter 8 for equations)

$$PFD_{avg} \text{ (sensors)} \quad = \quad .333 * (.01)^2(3/12)^2 \quad = \quad 0.000002$$
$$PFD_{avg} \text{ (valves and sol}^n) = \quad 2 * .333 * (.02)^2(3/12)^2 = \quad 0.000016$$
$$\text{Redundant PES:} = \quad = \quad 0.00001$$
$$PFD_{avg} \text{ (total)} \quad = \quad 0.000028$$

The maximum value required to SIL 3 is .001

This system therefore satisfies the safety requirements.

Nuisance trip calculations:

All transmitters and valves must be included in nuisance trip calculations.

$$MTTF_{sp} \text{ (sensors)} \quad = \quad 1/(4 * .01) = \quad 12.5 \text{ yrs.}$$
$$MTTF_{sp} \text{ (valves and sol}^n) = \quad 1/(4 * .1) \quad = \quad 2.5 \text{ yrs.}$$
$$MTTF_{sp} \text{ (PES)} \quad = \quad 1/(.01) \quad = \quad 100 \text{ yrs.}$$
$$MTTF_{sp} \text{ (total)} \quad = \quad 2.2 \text{ years}$$

A nuisance trip is expected to occur, on average, every 2.2 years.

This is also acceptable.

16.9 Perform Detailed Design

The detailed design has to comply with the safety requirement specifications and the conceptual design. For this application, the following documentation should be provided as part of the detailed design packages:

1. Results and recommendations of hazard studies.

2. All PFD_{avg} and MTTF calculations.

3. Process and instrumentation diagrams.

4. Instrument index.

5. Specification sheets for PES, transmitters, and valves

6. Loop diagrams.

7. Cause-and-effect matrix.

8. Drawings to indicate the locations of all major equipment and controls.

9. PES system configuration, I/O listings, and ladder program listings.

10. Junction box and cabinet connection diagrams.

11. Power panel schedules.

12. Pneumatic system tubing diagrams.

13. Spare parts list.

14. Vendor standard equipment documentation, including specifications, installation requirements, and operating and maintenance manuals.

15. Checkout forms and procedures.

16. Online test procedures.

16.10   Install the SIS, Commission It, and Perform Pre-start-up Tests

The following are the key activities which have to be completed as part of the installation, commissioning, and pre-start-up tests.

- Factory acceptance test (FAT) of the logic system. The responsibilities and details of the test are to be included as part of the detailed design activities. Support personnel are to witness the test.

- Field installation of the complete system. The field installation has to conform with the installation drawings that are included in the design package.

- Device functional checkout forms have to be completed and included with the installation scope of work (see Chapter 11).

- Pre-start-up acceptance test (PSAT). Because of the simple logic, the cause-and-effect diagram should be adequate for the PSAT.

16.11 Establish Operation and Maintenance Procedures

A typical trip test procedure is outlined below. The procedure covers the low pilot gas pressure trips for the furnace.

Operation:

If the pressure for either PT-7 or PT-24 drops below 5 psig, valves XV-30A, XV-30B, XV-31A, and XV-31B will close fully within three seconds.

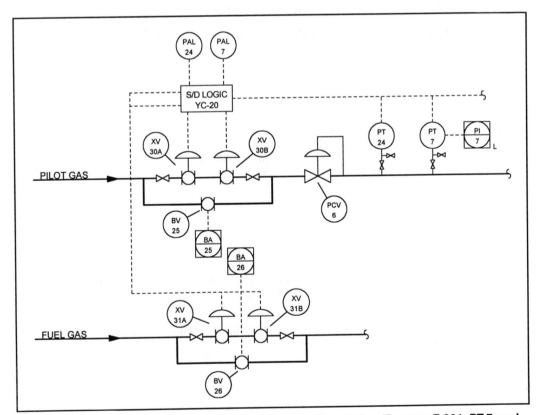

Figure 16-5   **Low Pilot Gas Pressure Trip System Description (Furnace F-001, PT-7, and PT-24)**

TRIP TEST PROCEDURE   **PROCEDURE #: XX-1**

1.0   <u>PURPOSE</u>

To test the low pilot gas pressure trips for furnace F-001 and transmitters PT-7 and PT-24.

2.0   <u>RESPONSIBILITY FOR TEST</u>

The furnace operator is responsible for completing the test. The resident instrument technician accompanies the operator during the test.

3.0   <u>TOOLS AND OTHER ITEMS REQUIRED</u>

Portable radio. No special tools are required to complete this test.

4.0   <u>TRIP CHECK FREQUENCY</u>

This test has to be completed every three months.

5.0   <u>HAZARDS</u>

Failure to open the bypass valves for the pilot gas or the fuel gas valves will trip the furnace and can create a furnace fire or explosion.

Failure to close the bypass valves fully upon completing the test can disable the furnace shutdown features

6.0   REFERENCE INFORMATION

The following documentation should be readily available to assist in rectifying any problems: loop drawings, instrument spec sheets, and shutdown system description.

7.0   DETAILED STEP-BY-STEP TEST PROCEDURE

Refer to Figure 16-5, which shows a trip system description.

RESULTS/COMMENTS/DEFICIENCIES

Step #	Action	Check OK
1	Advise all operating unit personnel that the test is about to commence.	
2	Open pilot gas bypass valve BV-25 fully. Verify that alarm BA-25 in the DCS came in. Acknowledge alarm.	
3	Open fuel gas bypass valve BV-26 fully. Verify that alarm BA-26 in the DCS came in. Acknowledge alarm.	
4	Get confirmation that furnace operation is still stable.	
5	Close the process isolation valve for PT-7 and slowly vent the signal to the transmitter. The two pilot gas trip valves and two fuel gas trip valves should trip at 5 psig. Record the trip pressure and verify that all four valves have closed fully and that the speed of operation is < 3 seconds.	
6	Verify that the following alarm was activated in the main control room: PAL-7 on the hardwired annunciator and in the DCS. Acknowledge alarm.	
7	Repressure PT-7.	
8	Alarms PAL-7 should disappear. Verify.	
9	Reset solenoids for four trip valves. Valves should reopen instantly. Verify.	
10	Repeat steps 5, 6, 7, 8, and 9 for PT-24.	
11	Close pilot gas bypass valve BV-25 fully. Verify that alarm BA-25 in DCS disappeared.	
12	Close fuel gas bypass valve BV-26 fully. Verify that alarm BA-25 in the DCS disappeared.	
13	Advise all operating personnel that the trip test for PT-7 and PT-24 is complete.	
14	Complete, sign, and send the completed documentation to the plant engineer for review and filing.	

TRIP CHECK COMPLETED BY _____   **DATE:_____**

Summary

The fired heater case study in this chapter shows how the techniques outlined in Chapters 1 to 15 can be used to specify, design, install, commission, and maintain a typical safety shutdown system. The following sections of the life cycle model are addressed for the application:

- define target SIL

- develop safety requirement specification (SRS)

- perform SIS conceptual design

- life cycle costs

- verify that the SIS conceptual design meets the SIL

- perform detailed design

- install the SIS, commission it, and perform pre-start-up tests

- establish operation and maintenance procedures

References

1. ANSI/ISA, Application of Safety Instrumented Systems for the Process Industries, ANSI/ISA S84.01-1996 (Research Triangle Park, NC: ISA, 1996), ISBN 1-55617-590-6.

INDEX